Business Process Modelling

Springer

Berlin
Heidelberg
New York
Barcelona
Budapest
Hong Kong
London
Milan
Paris
Santa Clara
Singapore
Tokyo

Bernd Scholz-Reiter
Eberhard Stickel (Eds.)

Business Process Modelling

With 117 Figures
and 12 Tables

 Springer

Professor Dr. Bernd Scholz-Reiter
Brandenburg Technical University of Cottbus
Industrial Information Systems
P. O. Box 101344
03013 Cottbus, Germany

Professor Dr. Eberhard Stickel
Europe University Viadrina
Department of Information Systems
P. O. Box 776
15207 Frankfurt/Oder, Germany

Business process modelling

Cataloging-in-Publication Data applied for

Die Deutsche Bibliothek - CIP-Einheitsaufnahme

Business process modelling / Bernd Scholz-Reiter ; Eberhard
Stickel (ed.). - Berlin ; Heidelberg ; New York ; Barcelona ;
Budapest ; Hong Kong ; London ; Milan ; Paris ; Santa Clara ;
Singapore ; Tokyo : Springer, 1996
    ISBN 3-540-61707-8
NE: Scholz-Reiter, Bernd [Hrsg.]

ISBN 3-540-61707-8 Springer-Verlag Berlin Heidelberg New York Tokyo

© Springer-Verlag Berlin · Heidelberg 1996
Printed in Germany

SPIN 10551142   42/2202-5 4 3 2 1 0 – Printed on acid-free paper

Preface

Today information is considered to be a valuable company resource that has to be planned, coordinated and documented. Business processes need to be treated in the same way. Moreover, significant potential for improvement may lie in optimizing these processes. As a consequence information should be collected and suitably modelled and business processes need to be analyzed and redesigned.

Business modelling usually deals with different views of a company. The data view is concerned with information structures, the functional view analyzes tasks and procedures, the process view links different functions by analyzing workflows and events triggering them, while the organizational view is concerned with the structure of the company under investigation. These views should not however be treated in an isolated manner. The key to success is to deal with the interaction between them. Hence, business modelling should provide an integrated view of all relevant informational, functional, organizational and workflow issues. One of the most significant problems results from the complexity of the modelling task. Modelling tools are required and are to be used in combination with other close methodologies and applications. Suitable generalizations and/or new methodologies need to be developed.

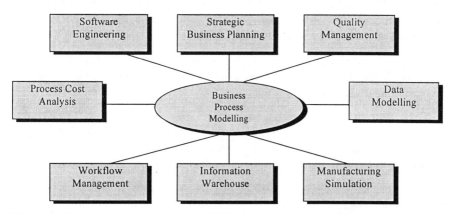

Fig. 1: Application Areas in Interaction with Business Process Modelling

Improvement and optimization of business processes calls for methodologies to value existing processes and restructure steps. Classical capital budgeting techniques are not directly applicable since returns and expenses are usually difficult to calculate in monetary terms. Existing applications need to be combined in order to generate support to the whole workflow that is initiated.

The goal of this book is to provide a forum for the exchange of new ideas in the field of business process modelling. The state of the art in business process modelling is reviewed, unsolved problems are identified and possible solutions are given. This volume is divided into the following chapters:
1. Interaction between Business Process Modelling and Workflow Management
2. Business Process Modelling for Engineering Processes
3. Strategic Business Planning
4. Business Process Modelling - Experiences
5. Methodological and Technical Aspects of Business Process Modelling.

In the first chapter Zukunft and Rump show how business process modelling can be used to derive workflow management concepts. Krallmann and Derszteler present an integrated approach to modelling, executing and monitoring workflow-based processes. Wieczerzycki's paper is also concerned with process modelling and workflow execution. Schmidt analyzes scheduling problems occurring in workflow management. Business process management systems are introduced by Karagiannis.

The support of engineering processes using the World Wide Web is analyzed by Grabowski et al. in the second chapter. The product development process is an example of so-called open processes. Modelling issues, as well as a suitable tool for modelling product development processes are discussed by Berndes and Stanke. Scholz-Reiter and Bastian examine groupware-based collaboration for distributed modelling of manufacturing systems.

Kamel and Lesca discuss Strategic Probe: A support-structured scanning information. Dietz and Mulder present a tool-supported method to realize strategic reengineering objectives.

The fourth chapter is devoted to practical experiences. Janson and Wrycza investigate the role of information technology during macroeconomic transformations in Poland. Chaffey discusses the application of business modelling techniques in an office automation-based process improvement programme.

The fifth chapter discusses methodological and technical issues of business process modelling. Yu investigates coordination-based approaches for modelling workflows. Rohloff presents an object-oriented approach. The reuse of reference process building blocks in business process reengineering is examined by Lang, Taumann and Bodendorf. Mehler-Bicher presents an architecture for a management support system that draws on object-orientation and business process modelling techniques. Finally Giaglis and Paul discuss the use of simulation techniques for business process redesign.

The editors would like to thank the members of the editorial committee and the numerous reviewers for their help in evaluating the high number of submissions for this volume.

Cottbus, Frankfurt (Oder) - Germany

July 1996

Bernd Scholz-Reiter, Eberhard Stickel

Editorial Committee

Dimitris Karagiannis (Austria)

Guy Doumeingts (France)

Otto Rauh (Germany)

Bernd Scholz-Reiter (Germany)

Elmar J. Sinz (Germany)

Eberhard Stickel (Germany)

Wolffried Stucky (Germany)

Business Process Modelling

Preface..V

B. Scholz-Reiter, E. Stickel

Part 1 Business Process Modelling and Workflow Management

From Business Process Modelling to Workflow Management: An Integrated Approach..3

O. Zukunft, F. Rump

Workflow Management Cycle - An Integrated Approach to the Modelling, Execution, and Monitoring of Workflow-Based Processes 23

H. Krallmann, G. Derszteler

Process Modelling and Execution in Workflow Management Systems by Event-Driven Versioning..43

W. Wieczerzycki

Scheduling Models for Workflow Management...67

G. Schmidt

Introduction to Business Process Management Systems Concepts........................81

D. Karagiannis, S. Junginger, R. Strobl

Part 2 Business Process Modelling and Engineering Processes

Implementation of Information Systems Supporting Engineering Processes Based on World Wide Web 109

H. Grabowski, M. Furrer, D. Renner, C. Schmid

Business Process Management for Open Processes: Method and Tool to Support Product Development Processes 128

St. Berndes, A. Stanke, K. Wörner

Distributed Co-operative Modelling of Production Systems 147

B. Scholz-Reiter, D. Bastian

Part 3 Strategic Business Planning

A Proposal Approach for Strategic Probe: A Scanning Information Support......175

K. Rouibah, H. Lesca

Integrating the Strategic and Technical Approach to Business Process Engineering 188

J. L. G. Dietz, H. B. F. Mulder

Part 4 Business Process Modelling - Experiences

Information Technology as an Enabler of Business Processes Designing During Macroeconomic Transformation 207

M. A. Janson, St. Wrycza

Integration of IT Strategy, Process Analysis and System Implementation in an Office Automation Based Process Improvement Programme...............................218

D. J. Chaffey

Part 5 Methodological and Technical Aspects of Business Process Modelling

A Coordination-based Approach for Modelling Office Workflow......................235

L. Yu

An Object Oriented Approach to Business Process Modelling............................251

M. Rohloff

Business Process Reengineering with Reusable Reference Process Building Blocks...265

K. Lang, W. Taumann, F. Bodendorf

An Object-Oriented and Business Process-Based Meta Model of an Architecture for Management Support Systems 291

A. Mehler-Bicher

It's Time to Engineer Re-engineering: Investigating the Potential of Simulation Modelling for Business Process Redesign 313

G. M. Giaglis, R. J. Paul

Part 1 Business Process Modelling and Workflow Management

From Business Process Modelling to Workflow Management: An Integrated Approach

O. Zukunft, F. Rump

Workflow Management Cycle - An Integrated Approach to the Modelling, Execution, and Monitoring of Workflow-Based Processes

H. Krallmann, G. Derszteler

Process Modelling and Execution in Workflow Management Systems by Event-Driven Versioning

W. Wieczerzycki

Scheduling Models for Workflow Management

G. Schmidt

Introduction to Business Process Management Systems Concepts

D. Karagiannis, S. Junginger, R. Strobl

From Business Process Modelling to Workflow Management: An Integrated Approach

Olaf Zukunft and Frank Rump

Universität Oldenburg, Fachbereich Informatik, Escherweg 2, D-26121 Oldenburg, Germany, {zukunft, rump}@informatik.uni-oldenburg.de

Abstract. Workflow management systems are used to support the computer based execution of processes in an enterprise. They require a description of these processes as well as an engine that implements the execution environment. Currently, workflow management systems have to introduce some new languages for the description of workflows in order to allow an execution of processes.

In this paper, we describe an approach which uses an existing and well accepted notation for the description of business processes as a workflow description language. This notation is already in wide use because it forms the basis for customising of the SAP R/3 system. In our approach, we reuse existing business process descriptions as input for our workflow management system. For this, we have realised a design component that allows a reengineering of existing business processes as well a design of new processes. These descriptions are mapped onto a workflow management engine built using an active database system. It turns out that this reuse of legacy business process descriptions is a feasible approach to workflow management.

1 Introduction

Recently, the terms "business processes" and "workflow management" have received a lot of attention. They are both used when talking about the automation of complex processes in office or industry. The term "business process modelling" is preferred when people concentrate on the analysis and design of processes, while "workflow management" is generally used when people also strive for supporting the execution of the formerly analysed processes through information technology. In this paper, we propose a novel and integrated approach to the problem of workflow management. Up to now, existing proposals have concentrated either exclusively on one part (modelling or execution) of workflow management or have introduced some new language constructs for business process modelling. Our approach allows to reuse an established language for business process modelling and combines it with a verification and validation of the modelled processes and with a reliable execution of these processes on database-centred systems. Hence, we are able to integrate "legacy" business process descriptions into a workflow management system and to implement the modelled processes in a distributed environment. This evolutionary approach is a critical feature for the successful application of workflow management systems. The EPC business process modelling language we (re)use forms the basis for the description of the SAP R/3 system. Enterprises using the SAP system are forced to model their business processes using EPCs in order to customise SAP R/3 (Keller 1995). Hence, there already exists a lot of knowledge and a lot of business process descriptions in these enterprises.

To implement the execution of workflows, we have to construct an engine which is capable of supporting the reliable execution of processes, handling multiple concurrently acting users, and which allows them to use a distributed system. The basic differences between the languages used for business process modelling and for the workflow engine are that they are used
1. by different people (business process versus workflow designers)
2. for different purposes (business administration versus workflow automation), and that
3. the later one has to be executable.

Ideally, we should be able to implement a compiler from the higher level BP modelling language to the language of the workflow engine. However, this is not possible in the general case because at the execution level we need additional information about the information technology infrastructure which can only be supplied by external sources and which the business process modeller should not be aware of. Hence, we argue that the mapping of business processes to a workflow engine can only be done semi-automatically and will always require some interaction with a workflow designer supplying additional information.

In this paper, we show how to perform workflow management using a variant of the EPC-model as the description language and an active database system as the enabling technology for the construction of a workflow engine. For this, the rest of this paper is organised as follows. We start by defining the basic concepts of business processes (BP) and workflow management systems (WFMS) as used in this paper. In section 3, we describe our approach to workflow design and analysis using the event-driven process chain (EPC). Afterwards, we discuss how to execute workflows described through business process modelling languages like EPC in section 4. The design and implementation of an integrated system supporting both the early phases of workflow modelling as well as workflow execution is subject of section 5. Related work is described in section 6. Finally, we present our conclusions and future work in section 7.

2 Preliminaries

In order to be able to describe business process and workflow management issues, we need a common understanding of what these and some related basic terms mean. Unfortunately, they are often used in the literature in different contexts with a different meaning.

In this paper, we use the term *business processes* when we talk about high-level descriptions of organisation's activities. These are used to describe market-oriented aspects like the satisfaction of customer needs. They received attention through *business process reengineering* (Hammer & Champy 1993), which is a method for the optimisation of the business processes by explicit redesign. Through the analysis of business processes, you obtain information where an optimisation of the processes is useful.

Workflows describe the processes of an enterprise on a lower level with focus on the execution of the processes. Thus, workflows are a collection of tasks organised to accomplish some business process. Workflows are divided in ad-hoc and administrative workflows (Georgakopoulos, Hornick & Sheth 1995). Ad-hoc workflows perform processes whose structure can not be predicted. The order of activities forming ad-hoc processes is defined only at runtime. These processes are usually supported by groupware systems. Administrative workflows have a predefined structure which is instantiated each time the workflow is performed. Thus, it is possible to support the process execution by informing the workflow participants about tasks to be done and by providing the information needed for the execution of a task.

Workflow management involves the (re)design and the (re)implementation of workflows as the needs and the goals of an enterprise change. A workflow management system (Workflow Management Coalition 1994) supports the computer based workflow management. That part of the WFMS providing the run time execution environment is called the *workflow engine*. This engine has to know the workflow specifications, the current instances of workflows, the information and organisational structure, and the available technology.

3 Workflow Design

In the area of business process modelling, most enterprises document their processes in an informal way only. This satisfies the need for documentation, but there is a big gap to workflow description languages if the computer based execution of the processes is required, too. Thus, if you want to support the business processes through a WFMS, you have to redefine the workflows from scratch and lose the connection to your previously described business processes.

To prevent this, we follow another approach by reusing an established, but informal model for the modelling of business processes, which is afterwards formalised and extended to fulfil the requirements for the description of workflows. Depending on the detail of description, we talk about business processes and workflows, respectively. For both, we offer different methods for analysing the models.

In the next section, we introduce the widely used EPC-model, which forms the base of our approach. Section 3.2 describes the EPC*-model, which augments the EPC-model with formal syntax and semantics, as well as a methodology for reengineering existing EPCs.

3.1 The EPC-model

The architecture of integrated information systems (Scheer 1992) splits the total view of an enterprise into four different views: *data view*, *function view*, *organisation view*, and *process view*. The process view integrates some parts of the other views because it describes the dynamics of an enterprise using information from the static views. This view documents the business processes of an enterprise.

For the description of the static views, established models can be used, e.g. the ER-model for the data view. Because of the requirement to reflect organisational, functional, data oriented and dynamic aspects in the process view, the event-driven process chain model (EPCM, (Keller, Nüttgens & Scheer 1992)) was developed.

The EPCM uses the following graphical symbols:

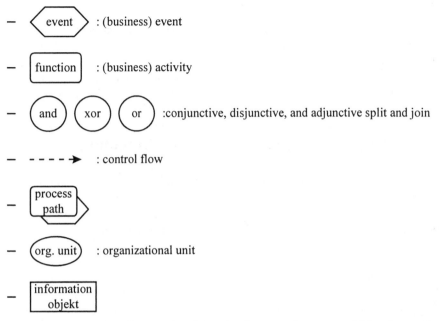

- **event** An event reflects a signal in a business environment which triggers the execution of a function. Thus, it is a passive element in the process.
- **function** A function reflects the active elements of a process in the EPC-model and describes an activity. It is executed either by a person or automatically. Linked with it is information describing which input data is used and which output data is produced. Furthermore, you can connect a function with an organisational unit to define the responsibility for the execution of the function.
- **split/join** A split or join describes the logic of the control flow. The AND-split executes parallel paths, which can be synchronised by an AND-join. Of an XOR-split exact one following activity is started. Thus, a synchronisation of an XOR-join is not needed. The OR-split corresponds to a "don't care" split, thus all combinations of the following paths are possible. Depending on the specific case, an OR-join must synchronise the incoming paths.
- **process path** Process paths are connections to other EPCs to divide business processes into various EPCs.
- **organisational unit** This is connected with a function to define the responsibility for the execution.
- **information object** This describes which data is needed for the execution of a function and which data is produced by the function.

The initial goal of the EPCM was to create a simple documentation of the business processes of an enterprise. It is used to inform the persons involved in the process about their tasks. An execution of the workflows based on the defined schemes is not intended. Thus, the EPCM contains some shortcomings, which must be removed to enable the execution of EPCs.

The weakest point of the EPC-model is the missing formal syntax and semantics. A formally defined syntax and semantics is not required for documentation of the processes, but is essential for analysis and execution. Furthermore, the EPC-model does not offer a possibility to specify the data flow, the conditions of the control flow and the functions more detailed (Rump 1995).

3.2 Reengineering EPCs for Enabling Execution

To overcome the shortcomings described above, we have defined an extension of the EPCM to enable the execution of workflows. This extension is called the EPC*-model, and will be introduced in the following. Furthermore, we describe methods for analysing schemes and for reengineering schemes of the EPC-model.

3.2.1 The EPC*-model

An EPC* is a graph $G_{EPC*} = (K_N, K_E)$ with the set of nodes $K_n = E \cup F \cup J \cup P \cup I \cup O$ and the set of edges $K_e = K_K \cup K_D \cup K_O$.

- E is a non-empty set of events.
- F is a non-empty set of functions.
- J is a set of splits and joins.
- P is a set of process paths.
- I is a set of input and output containers.
- O is a not empty set of organisational units.
- $\forall X, Y \in \{E, F, J, P, I, O\}: X \neq Y \Rightarrow X \cap Y = \emptyset$
- $K_K \subseteq (K_n \setminus (I \cup O)) \times (K_n \setminus (I \cup O))$ describes the control flow.
- $K_D \subseteq (I \cup F) \times (I \cup F)$ describes the data flow between input/output containers and the connection between containers and functions.
- $K_O \subseteq O \times F$ specifies the connection between organisational units and functions.
- $\forall X, Y \in \{K_K, K_D, K_O\}: X \neq Y \Rightarrow X \cap Y = \emptyset$
- The map $cond: K_K \rightarrow COND$ assigns a condition to an edge out of a set of allowed conditions (e.g. for XOR-splits).

- The map $script: F \rightarrow SCRIPT$ assigns a script to a function in a given language. This can be an iterative program or a call of another system (e.g. word processor).

The structure of an EPC* is defined using these graphs. The set of syntactical correct EPC*s must furthermore fulfil some additional properties, which are not listed here due to space limitations. For the main part of the EPC*-model we have defined an operational semantics which allows to employ advanced analysis techniques.

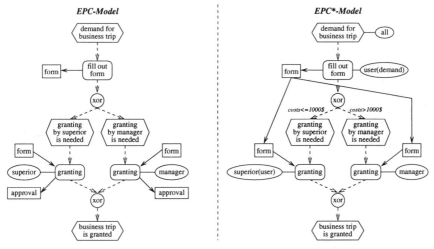

Fig. 1: Comparison of EPC and EPC*

In figure 1, an EPC for the process of granting a business trip is shown. If a person wants to make a trip, a form is filled with the needed information. If the costs are less than 1000$, the superior of that person can grant the trip, else the manager of the enterprise has to grant the trip. On the right side, the EPC* for the same business process is shown. This description corresponds to the syntax of the EPC*-model, thus it can be used for further analysis of the process. The start event "demand for business trip" is connected with an organisational unit ("all") to specify who can start this business process. The predefined value "all" means that all persons in the enterprise can make a business trip. Depending on the person who initiates the process, the functions of the process are executed. The function "fill out form" is for example performed by the person who initiated the start event. Another difference in the EPC*-model is the explicit specification of the data flow by connecting the input and output containers of functions (see information object "form"). Furthermore, the conditions of XOR- and OR-splits are explicitly defined with information from the previous function (see condition "costs<=1000$", where costs is an attribute of the information object "form").

3.2.2 Analysis of EPC*s

By analysis of business processes we mean the validation and verification (Bröckers & Gruhn 1993). Validation allows to show properties of a specific instance of an EPC* through simulation. You can get for example an impression of the dynamic behaviour of the process by visualisation of an EPC*. If you assign information like costs and a period of time to functions, then the costs and the time period of the whole process as well as the capacity utilisation of persons in an enterprise can be calculated. These information is necessary to support *business process reengineering (BPR)*. Furthermore, a simulation of an EPC* with test data is possible.

The verification shows instead properties of all possible process instances described by an EPC*. Thereby it is possible to find out properties like reachability or to find OR-Joins which are always used as AND- or XOR-Joins. Because of the use of splits and joins, it's quite easy[1] to make a mistake in a scheme by introducing inconsistencies in the process logic. By verification you can find such inconsistencies, which is a sort of quality check of modelled business processes.

Fig. 2: Inconsistency in process logic

In figure 2 an inconsistency in the process logic is shown. An AND-split is "synchronised" by an XOR-join, which leads to an error, if both incoming paths of the XOR-join are executed. In this case, the verification would warn that the XOR-join must be replaced by an OR-join.

---

[1] The R/3 reference model describes by approx. 800 EPCs all business processes which are supported by the SAP R/3 system. In this model we found many inconsistencies in the process logic which make a simulation impossible.

3.2.3 How does Reengineering work in this context?

Currently, there already exist many descriptions of business processes using the EPC-model. Most of them have been created for documentation purposes only. However, the execution of these processes is more and more often desired. Doing this the traditional way requires to start from scratch by creating a workflow specification defining the processes with the description language of a workflow management system. This is not acceptable if the designer is not supported in reusing the previously defined business processes.

To prevent the multiple definition of identical processes, our approach offers a continuous methodology for reengineering and executing business processes on the basis of the EPC*-model. We support an easy transformation of legacy business process descriptions written using the EPC-model into a formally defined and hence automatically processable language. As described above, our language is close to the well known language EPC but forms nevertheless a strong basis for the execution of processes. To achieve this, our methodology is divided into the following steps:

1. **syntax check:** Because of the missing syntax definition of the EPC-model, you have to run a syntax check on the existing schemes to make further improvements possible.
2. **verification:** After the syntax check, the schemes are verified to find possible inconsistencies in the process logic.
3. **validation I:** This step allows a visualisation of the processes, a cost and a time analysis. Depending on the results, an improvement of the schemes may be useful.
4. **specification:** In this step, the workflow designer augments the schemes with information needed for the execution of the workflows. He has to describe the conditions for the splits, the functions and the data flow.
5. **validation II:** Because of the further specified workflows, a validation based on test data is possible.
6. **execution:** After executing the previous steps the workflows are now executed using our workflow engine.

The steps syntax check, verification, and validation I are carried out by the business process designer. A workflow designer, which has a sound knowledge of the EPC*-model, is responsible for the following steps of specification and validation II. The execution is initiated and supervised by the workflow administrator.

4 Workflow Realisation

After the design and analysis of business processes, a WFMS has to offer a technical subsystem which is capable of supporting the execution of workflows.

The type of subsystem which has to be offered depends on the kind of workflow we want to support. For ad-hoc workflows, we basically need a sophisticated tool for supporting communication between the workflow participants. For structured and partly structured workflows, we are able to offer more support for the execution of workflows because there we can reuse the formerly created information about the structure of the workflows for the execution. In the following, we will focus our discussion on those structured workflows that are described using the EPC*-model.

4.1 Active Database Technology for Workflow Engines

We have decided to use the technology of active database systems for implementing our workflow engine. Active database systems (Dayal et al. 1988) integrate an event-based rule system into the DBMS and use ECA-rules for this purpose. Hence, they offer a language which allows to model both the data and the transactions within the database system. Furthermore, active database systems extend the notion of transactions onto the dynamic part of the application through their realisation as rules within the database system. This strongly supports the creation of a workflow engine which is capable of handling failures in the execution of workflows and which manages a concurrent access of multiple users to the workflow system. Additionally, modern active database systems use non-standard transactions models (Buchmann, Özsu, Hornick, Georgakopoulos & Manola 1992) to offer functionality for the rule execution. These models do already support some of the advanced requirements of workflow engines on transactions, namely the higher structure through a nesting of transactions, the open system characteristic through compensating transactions (open nested transactions) and intertransactional relationships like causality through coupling modes (Branding, Buchmann, Kudrass & Zimmermann 1993). Regarding the requirement of being close to workflow modelling languages, active database systems offer not only the capabilities of classical database systems for storing structured data, but also a rule language which allows to implement the control structures used by workflow modelling languages. While the organisation and data aspects of the workflow model language are mapped onto the implemented data model (i.e. relations or classes), the activities and control flow are mapped onto the rule language. Because of space limitations, we will focus in the following on the description of the activities.

Regarding the control flow structures, we have to support at least the constructs defined by the workflow management coalition (WFMC), namely joins, splits, a sequence and an iteration. This is done using the following schema (see appendix A for details):

- AND-Join

  For every AND-join, we generate one rule which is triggered by the conjunction of signals raised at the end of the joining activities.

- (X)OR-Join

  In the EPC*-model, we distinguish between an exclusive and a non-exclusive OR-join. An exclusive join synchronises an exclusive OR-split and is triggered by the occurrence of the end of the only split activity. For every exclusive OR-join, we generate one rule which is triggered by the disjunction of signals raised at the end of the joining activities. A non-exclusive OR-join requires access to data describing the workflow on an instance specific level. It is used to synchronise a "don't care" OR-split, i.e. a split where a previously unknown number of activities are started. To implement the OR-join, we have to know which activities have been split. Since this can only be determined at runtime, we generate a rule which uses a conjunction of all activities, whether started or not. For all non-running activities, an event is artificially generated to fulfil the complex event.

- AND-Split

  For every AND-split, we generate a set of rules so that every element of the rule set executes one of the actions in the split. Consequently, the WFMS is able to execute the split activities concurrently. Note that the AND-split does not require a condition to be associated with the splitting of activities.

- (X)OR-Split

  The OR-split is usually used to initiate an action which is dependent on some outcome of formerly executed actions. Therefore, a condition is associated with any of the split actions. In the EPC*-model, we distinguish between exclusive and non-exclusive OR-splits. For both, we generate a set of rules which initiate the execution of the split activities. The semantic difference between them is that only one of the actions specified in an exclusive OR-split is allowed to be executed. Although it is the responsibility of the workflow designer to assure that the conditions of the actions are mutually exclusive, we use the detached causally dependent exclusive coupling mode introduced in Branding et al. (1993) to guarantee this behaviour. For non-exclusive OR-splits, we additionally generate a rule raising the termination event if the activity is not started. This is required for correct synchronisation of the OR-join.

- Iteration

  For an iteration, we generate an additional rule triggering the action to be re-executed.

- Sequence

  To realise a sequence, we generate one rule executing the sequence in the action of the rule. This rule is triggered by the event leading to the execution of the first action in the sequence.

Through this explicit modelling of control within the database system, the workflow execution is largely independent of the business process modelling language. Instead of reprogramming the mapping from business process modelling to workflow execution, we only have to redefine the rules for every modelling language to be used. Beyond this flexibility, we are furthermore able to use optimisation techniques and tool support directly on the rule level.

While the discussion above shows that active database systems alone can offer a strong basis for the implementation of the workflow engine, we nevertheless want to stress that its construction requires more than an active database system can offer. First, we also have to integrate components other than an active DBMS for building the WFMS. This includes a user interface which can be the source of events triggering the execution of workflows and legacy applications which are used for the execution of actions. Therefore, user interface events have to be usable in complex events within the rule language and arbitrary foreign programs must be callable in its action part. Secondly, existing active database systems do not explicitly support a multi-user capable rule system. This includes for example the visibility of events raised by one user for others, where the users have different privileges as well as the switching between different rule sets for different workflows within one active DBMS.

5 Design and Implementation

In this section, we describe the design and implementation of the Waterloo[2] workflow management toolbox. Waterloo supports both the early phases of business process design as well as the later phases of workflow execution.

---

[2] Please note that Waterloo was a defeat for Napoleon only!

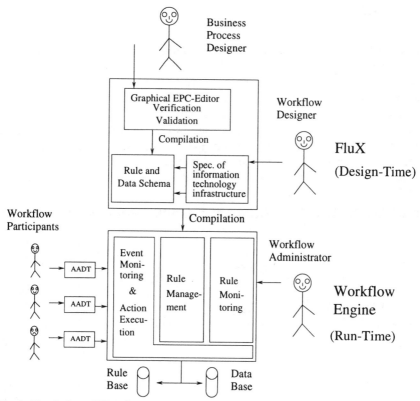

Business
Process
Designer

Graphical EPC-Editor
Verification
Validation

Workflow
Designer

Compilation

FluX

Rule and
Data Schema

Spec. of
information
technology
infrastructure

(Design-Time)

Compilation

Workflow
Participants

Workflow
Administrator

Event
Moni-
toring
&
Action
Execu-
tion

Rule
Manage-
ment

Rule
Moni-
toring

AADT

AADT

AADT

Workflow
Engine

(Run-Time)

Rule
Base

Data
Base

Fig. 3: The design of Waterloo

The principal design of Waterloo is shown in figure 3. On the top level, we have a tool called FluX which supports the design phases of the workflow management. It is used by the business process designer for modelling and analysis of business process as well as by the workflow designer who augments the information entered by the business process designer with information about the information technology infrastructure like the actual layout of the network and the interface to foreign systems used in implementing the workflow. The process oriented view modelled by the business process designer is taken by a compiler and converted into a set of rules, relations and data, which forms together with the information provided by the workflow designer the input for the workflow engine. The combination of these two information sources with the generic workflow engine allows to create an executable system as shown at the bottom level of figure 3. This is implemented as an application of our active information system toolbox AIDE (Jasper 1994). The workflow participants who are finally responsible for executing the workflows are interacting with this active information system through graphical user interfaces. We will discuss both of these layers in the following.

5.1 FluX

FluX is our process definition tool for the modelling of business processes and the specification of workflows. Based on the EPC*-model, it supports both the modelling from scratch and the reengineering of given EPCs as described in section 3.2.3. Implemented using Tcl/Tk and C++, FluX uses a repository with an underlying SQL92-DBMS to store the business process descriptions. There, the EPC*s are grouped into projects, which are associated with the processes of an enterprise. The editor component supports the modelling of business processes and workflows through an adaptable graphical user interface (see figure 4). Furthermore, an analysis component covers the validation of business processes and the verification of business process properties.

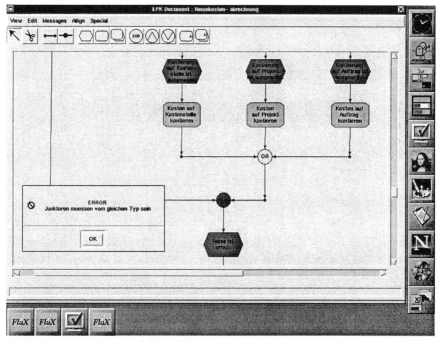

Fig. 4: A snapshot of the FluX user interface

The design of FluX allows an easy integration of other models and graph types. Thus, different graphical languages for the specification of business processes can be modelled with FluX and links between schemes of different models are possible. If a formal mapping of the models onto EPC* is supplied, it is possible to analyse business processes described using different models.

5.2 Workflow Engine

While FluX is used at design time, we also provide a tool offering the user support at runtime of workflows. This is done by a workflow engine. We have used the AIDE (active information development environment) toolbox for the construction of this runtime system. AIDE supports the integration of arbitrary subsystems through the concept of active abstract datatypes (AADT). In addition to classical abstract datatypes, an AADT provides a set of events which are raised by the subsystem encapsulated in the AADT. We have used a Tcl/Tk-AADT for the construction of the user interface and an SQL2-AADT for the construction of the database interface. Hence, we use the functionality of relational database systems and the Tcl/Tk-environment in the execution of those active rules that implement the activities and the control flow within the workflow engine. Furthermore, the concept of AADT provides a transparency of distribution. Both the SQL2- and the Tcl/Tk-AADT support distribution of functionality.

For the coordination of the activities, we use the conceptual mapping schema described in section 4. The static aspects of the design are mapped onto a relational schema. This schema is independent of the actual workflow design and only dependent on the workflow design language used. Data transferred between the activities is mapped onto parameters of those events initiating the next activity. These as well as the control structures are implemented using the rule language of our active database system. For the implementation of the activities, our rule language offers a computational complete sublanguage with imperative constructs. For fully automated actions, we are able to execute arbitrary foreign procedures. For semi-automated actions, we can request a user interaction. For the triggering of non-automated actions, our workflow engine computes the set of people capable of executing the activity and displays the activities on their worklist.

Since the area of business process modelling is a very dynamic one, we have decided not to build a close coupling between the workflow design and the runtime systems. Instead, we have tried to build the workflow engine independent of the language used for the specification of workflows. The only assumptions made are that the workflow engine has access to an organisation model, to a description of the data transferred between activities, and that the control structures follow the model of the WFMC. To test this language independence, we have recently finished a compiler for the definition language used by the FlowMark system (Leymann & Altenhuber 1994) as an example of the feasibility of this approach.

Complementing the WFMC model for workflow design, we have designed an advanced time-model in the workflow engine which we are able to offer at the higher levels of business process and workflow design. This time model allows to specify start and finish of activities, either absolute or relative to some other event, and a maximal duration of activities. While specifying a point of time for the start of an activity is supported by most workflow engines, a specification of an end point requires a scheduling of activities. This can only be supported if the workflow engine uses techniques for a real-time execution of transactions (Jasper, Zukunft & Behrends 1995).

While the availability of the time model is an advantage of the decision to use an active DBS for realising the workflow engine, we have also experienced some problems during the implementation of our workflow engine with our active database system. These concern:

- Workflows are often instance specific workflows, i.e. different workflows of one type show different behaviour due to user interactions. This results in the requirement of supporting instance specific rules, unless we introduce a new type for each workflow.
- In order to visualise the status of a business process, we need access to the internal state of the active part of the DBMS. To the best of our knowledge, this has not been previously supported by any active DBMS.
- Our active DBMS does not support all coupling modes proposed in Branding et al. (1993). To change this, we need to modify the internals of the underlying transaction manager.

Summarising, the concept of using active database technology together with the possibility to integrate technical and human subsystems has shown to be a sound basis for the implementation of a workflow engine.

6 Related Work

Business process design and workflow management has been a topic of active research. The concept of EPC can be found in Keller et al. (1992). Opposed to our work, their approach does neither allow an analysis nor an execution of workflows. Both of these features are supported by a number of workflow management systems like FlowMark (Leymann & Altenhuber 1994), InConcert (Hsu (Editor) 1993) and ActionWorkflow (Medina-Mora,R., Winograd, T., Flores, R. & Flores, F. 1992). However, these systems all use some newly defined language for the modelling of business processes which is intended to be the input language for the WFMS. This differs significantly from our work in that we explicitly base our WFMS on existing business process descriptions. Using active database systems for workflow management has been first proposed by Bussler & Jablonski (1994) and later by Geppert, Kradolfer & Tombros (1995). However, they both concentrate on the technical issues of realising workflow management through active database systems only and do neither consider the business process modelling part nor allow an execution of processes based on existing process descriptions.

7 Conclusion and Future Work

In this paper, an integrated approach for the modelling, analysis, and execution of workflows based on existing business process descriptions has been proposed. We have shown how to reengineer process descriptions that are based on the EPC-model, and described a workflow engine able to execute augmented processes. Through the exploitation of knowledge existing in enterprises, it is possible to significantly reduce the effort required to introduce workflow management systems. To support this, a system for the design, analysis and execution of workflows based on our approach has been described. Both components, the design tool and the execution engine are implemented. The execution engine uses the technology of active database systems to support a reliable and flexible implementation of business processes. Deficits of active database systems used for this application concern the explicit support for multi-users in the rule system and a missing distribution of rule processing. However, our system is able to support workflow participants using a distributed system.

In the future, we plan to consider a multi workflow engine approach based on a distributed active database system. This would significantly enhance the robustness of the execution engine. Since workflow participants are more and more often using PC, we further plan to port our workflow client environment from UNIX to the Windows operating system. Additionally, we intend to integrate support for the realisation of ac-hoc workflows into our Waterloo system. However, it is not yet clear how this can be integrated into the EPC*-model. To integrate mobile users into our WFMS, we currently investigate extensions at both the modelling and the implementation level (Zukunft 1996). Finally, we want to gain additional application experiences by using Waterloo in the area of hospital information systems.

References

Branding, H., Buchmann, A., Kudrass, T. & Zimmermann, J. (1993), Rules in an open system: The REACH rule system, in *First Intl. Workshop on Rules in Database Systems*, Edinburgh.

Bröckers, A. & Gruhn, V. (1993), Computer-aided verification of software process model properties, in *Proceedings of the Fifth International Conference on Advanced Information Systems Engineering (CAiSE)*, pp.521-546.

Buchmann, A., Özsu, M. T., Hornick, M., Georgakopoulos, D. & Manola, F. A. (1992), A transaction model for active distributed object systems, in *Database Transaction Models for Advanced Applications* (ed. A.K. Elmagarmid), Morgan Kaufmann, San Mateo, CA, chapter5.

Bussler, C. & Jablonski, S. (1994), Implementing agent coordination for workflow management systems using active database systems, in *Proc. 4th Intl. Workshop on Research Issues in Data Engineering*, Houston, USA, pp.53-61.

Workflow Management Coalition (1994), *Glossary*.

Dayal, U., Blaustein, B., Buchmann, A., Chakravarthy, U., Hsu, M., Ledin, R., Mc-Carthy, D., Rosenthal, A., Sarin, S., Carey, M., Livny, M. & Jauhari, R. (1988), The HiPAC project: Combining active databases and timing constraints, *SIG-MOD Record* 17, 51-70.

Georgakopoulos, D., Hornick, M. & Sheth, A. (1995), An overview of workflow management: From process modelling to workflow automation infrastructure, *Distributed and Parallel Database Systems* 3(2), 119-153.

Geppert, A., Kradolfer, M. & Tombros, D. (1995), Realisation of cooperative agents using an active object-oriented database management system, in *Proc. 2nd. Intl. Workshop on Rules in Database Systems*. LNCS985, Springer, Athens, Greece, pp. 327-341.

Hammer, M. & Champy, J. (1993), *Reengineering the Corporation: A Manifesto for Business Revolution*, Breadly, London.

Hsu (Editor), M. (1993), Special issue on workflow and extended transaction systems, *Bulletin of the Technical Committee on Data Engineering*.

Jasper, H. (1994), Active databases for active repositories, in *Proc. 10. Intl. Conference on Data Engineering*, IEEE Press, Houston, USA, pp.375-384.

Jasper, H., Zukunft, O. & Behrends, H. (1995), Time issues in advanced workflow management applications of active databases, in *Proc. First Intl. Workshop on Active and Real-Time Database Systems*, Springer, Skövde, Sweden, pp. 65-81.

Keller, G. (1995), Creation of business processes with event-driven process chains (EPCs): A strategic challenge, in *SAPinfo - Business Reengineering*, SAP AG, Walldorf, pp. 8-13.

Keller, G., Nüttgens, M. & Scheer, A.-W. (1992), Semantische Prozeßmodellierung auf der Grundlage Ereignisgesteuerter Prozeßketten (EPK), in *Veröffentlichungen des Instituts fur Wirtschaftsinformatik* (ed. A.-W. Scheer), number 89, Saarbrücken.

Leymann, F. & Altenhuber, W. (1994), Managing business processes as an information resource, *IBM Systems Journal* 33(2),326-348.

Medina-Mora, R., Winograd, T., Flores, R. & Flores, F. (1992), The action workflow approach to workflow management technology, in *Proceedings of the ACM Conference on Computer Supported Cooperative Work (CSCW'92)*, Emerging technologies for cooperative work, ACM Press, Toronto, Ontario, pp.281-288.

Rump, F. (1995), Ereignisgesteuerte Prozeßketten zur formal fundierten Geschäfts-prozeßmodellierung, in *Informationssystem-Architekturen* (ed. E. Sinz), Vol.2, GI-Fachausschus 5.2, pp. 94-96.

Scheer, A.-W. (1992), *Architecture of Integrated Information Systems*, Springer.

Zukunft, O. (1996), Application partitioning in workflow management systems supporting mobile clients, in *Proc. International Workshop on Information Visualization & Mobile Computing*, Rostock, BRD.

A Rule schema for control structures

```
RULE AND_Join
ON (a1_completed AND a2_completed AND ... AND an_completed)
IF TRUE
DO Action()
END RULE


RULE XOR_Join
ON (a1_completed OR a2_completed OR ... OR an_completed)
IF TRUE
DO Action()
END RULE


RULE OR_Join                            //Non-exclusive join
ON ((a1_completed OR a1_notStarted) AND
    (a2_completed OR a2_notStarted) AND
    ...
    (an_completed ORan_notStarted))
IF TRUE
DO Action()
END RULE


RULE AND_Split_Part_N     // One rule per split activity
ON a1_completed
IF TRUE                   // No Condition
DO Action_N();
END RULE


RULE OR_Split_N_Part1     // One rule per split activity
ON a1_completed
IF condition_N   // Disjoint conditions for exclusive splits
DO Action_N
END RULE


RULE OR_Split_N_Part2     // One rule per split activity
ON a1_completed
IF NOT(condition_N)
DO raise_event(Action_N_notStarted); // For correct OR-Join
END RULE


RULE Iteration
```

```
ON    a_completed
IF    condition
DO    Action;      // Action is also triggered by another rule
END RULE

RULE Sequence
ON a1_completed
IF TRUE
DO Action_1();
   Action_2();
   ...
   Action_N();
END RULE
```

Workflow Management Cycle - An Integrated Approach to the Modelling, Execution, and Monitoring of Workflow Based Processes

Hermann Krallmann and Gérard Derszteler

Chair of Systems Analysis and EDP, Institute of Business Computing, Technical University of Berlin, Franklinstr. 28/29, D-10587 Berlin, Germany, krallm@cs.tu-berlin.de, gerard@sysana.cs.tu-berlin.de

Abstract. The modelling of effective and efficient process structures is important if reengineering projects are to be successful. However in addition to this, methodical and technical support for implementing and monitoring these processes is also becoming a key factor. This article presents a cyclic model for the holistic management of workflow based processes. The approach integrates common reengineering and workflow tools. The aim is continuous improvement of the corporation's flow structure.

1 Introduction

In the last few years understanding of the role of a company's organisation has dramatically changed. Previously the main purpose of a company's organisation was to establish competencies and responsibilities for all employees. The optimisation or even the explicit description of processes was a subordinate goal presented by the organisational design task.[1]

1.1 Business Process Reengineering

Since the beginning of the 1980s organisational structure is no longer seen to dominate flow structure.[2] The need for efficient and effective resource consumption was rising due to higher competitive pressure and so the main business activities were focused on all further considerations. With regard to this Porter's value chain model is well known. This model divides all main business activities into direct (primary) value generating activities and indirect (secondary) activities, which support the primary ones. In order to increase the competitive edge of a firm, all activities which do not belong to one of these groups must be eliminated.[3]

It took several years for the theoretical formulation of these new concepts to be widely adopted by industry. Since the beginning of the 1990s several new reorganisation approaches were developed, which are known by the term *business process reengineering* (BPR).[4] This process was accelerated predominantly by consulting firms.

BPR is a synonym for the fundamental renewal of corporate structures aiming to create customer-oriented business processes. In this way focus is placed on the core business processes, with no traditional organisational borders. Revolutionary alterations are often called for. "Reengineering, properly is the fundamental rethinking and radical redesign of business processes to achieve dramatic improvements in critical contemporary measures of performance such as cost,

---

[1] See for example (Frese, 1992), or (Kieser and Kubicek, 1992).
[2] (Gaitanides, 1983)
[3] See (Porter, 1992a), and (Porter, 1992b).
[4] (Davenport, 1993), and (Hammer and Champy, 1993) are two of the best-known publications in this field.

quality, service, and speed"[5]. These authors claim that such radical actions are justified by the enormous rewards, which cause a quantum-leap in productivity.

1.2 Computer Supported Cooperative Work and Workflow Management

Process orientation causes a reintegration of elementary work pieces and therefore partially retracts the very high specialisation of the employees. This results from the functional decomposition of business tasks in today's corporations. This job enlargement often connected with job enrichment is made possible by a new generation of software: the *computer supported cooperative work systems* (CSCW) also known as groupware systems.[6] Based on modern information technology (IT), consisting of corporation-wide client/server-systems with integrated legacy applications and personal computers on every desk, groupware systems bridge gap between spatial and temporal dimensions. This is essential for processes across a corporation with a minimum of job-breaks and ideally no medium-breaks. The broad spectrum of collaborative support consists of a variety of systems. These are often specialised for specific cooperation situations. The most widely recognised classification criteria are the previously mentioned dimensions room and time, and the structuring extent of the supported task.[7]

Workflow management systems (WFMS) are a particular kind of collaboration system which supports well-defined processes in highly-dispersed organisations.[8] A workflow management system is a computer based system, which controls the flow of work (business processes) between people depending on their role in an organisation. The flow of work is described by detailed modelled business transactions. Every business transaction consists of a set of logically ordered elementary functions, known as activities.[9] Significant improvements in productivity are possible with the use of WFMS. Information is no longer restricted to paper, and - particularly interesting from the customer's point of view - throughput time is dramatically reduced.

---

[5] (Hammer and Champy, 1993, p.12)
[6] See for example (Hasenkamp and Kirn and Syring, 1994), or (Dier and Lauterbacher, 1994).
[7] (Hasenkamp and Syring, 1994, p.19)
[8] (Götzer, 1995)
[9] (Karagiannis, 1994, p.110)

2 Holistic Process Management

The extremely high expectations presented in many articles, for the success of BPR projects are often not reached. Reasons for this are the radical demands for restructuring which are often unrealistic. On the whole no practical hints are given on how to implement them.[10] Radical renewal is one of the core elements of BPR. This approach is appropriate for establishing a fundamentally new technology or innovation, but cannot be used for regular application due to the great effort needed. Alongside these quantum leaps, continuous improvement of the process structures by means of many small single steps reflecting the corporation's learning curve is the preferable course of action.[11]

Österle identifies the evolutionary approach in contrast to the revolutionary method. The evolutionary advance is slower and does not produce fundamental improvements, but allows routine analysis. The improvements gathered in this way can be gradually implemented without affecting normal business.[12] The basic idea of continuous improvement comes from the production area, where measures such as suggestion systems and quality circles were developed. Alongside with other methods these belong to the field of *total quality management* (TQM).[13]

The evolutionary and revolutionary approaches do not exclude one another. On the contrary only their combination allows a continuous and long-term improvement process. Thus the graph illustrating the overall enterprise productivity has an appearance similar to a flight of stairs (e.g. a sharp rise followed by a period of no or little change). BPR projects carried out rarely but thoroughly lead to an evolutionary cycle in which the corporation remains until the next substantial innovation is implemented. Fig. 2.1 illustrates this relationship.

---

[10] (Gaitanides and Scholz and Vrohlings and Raster, 1994, p.V)

[11] (Porter, 1992a, p.35)

[12] (Österle, 1995, p.22)

[13] (Seghezzi, 1996, p.111)

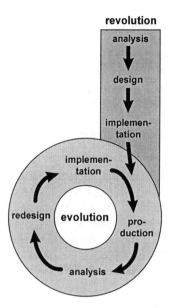

Fig. 2.1: Revolutionary and evolutionary approach to business process reengineering (Österle, 1995, p.23)

2.1 Process Management Methods

Within the framework of systematic process improvements there are other essential tasks besides process modelling. A detailed analysis of current process structures as a basis for following reorganisation steps and an efficient implementation of the new structures are also crucial for successful process management. The basic idea behind this is a plan-implement-analysis-cycle.

The main task in the planning stage is to develop new flow structures. A thorough analysis of the current state is the input data for this stage. The new process structures are systematically analysed before implementation. This task is supported by computer-based reengineering tools such as the ARIS-Toolset or Bonapart. With the help of these programs process models could easily be drawn. They also help by checking the model's consistency and efficiency by providing simulation and analysis modules. In addition to the process models, goals and objectives for the operative level, for example the throughput time of processes, are also defined in the planning phase. These objectives are later used for variance analysis.

WFMS are used to carry out the modelled processes. They control and guide execution of processes.[14] Data for all process instances can be gathered by certain systems and thus a detailed protocol can be created. This raw information is very important as a basic input into the evaluation phase. The main task for the feedback analysis is to compare the objectives and goals given by the modelling phase with the protocolled data of completed workflows. Information gathered about strengths and deficiencies of the current organisation structure is used for the planning period. Until now there has been very little computer support for this target-performance comparison.

The previously described cyclic execution of modelling, implementation, and analysis of processes is called *process management*, although various terms are used in literature. Heilmann describes the concept as Workflow Management, which she strictly differentiates from Workflow Management Systems. Karagiannis characterises the whole cycle as a Business Process Management System, whereas Deiters and Striemer call it a Business Process Engineering Life Cycle.[15] All models share - in rough - the same sequence of operations: process planning, implementation support, and ex-post analysis. This procedure is based on the classical feedback control circuit of cybernetics.[16]

2.2 Critical Points Regarding Existing Process Management Methods

Different approaches have been developed with regard to the idea of a process management cycle.[17] These methods usually have two things in common. Firstly the modelling phase is supported by reengineering tools. And secondly WFMS handle the process execution. However, maybe due to this, there are still problems with regard to the efficient application of these cycle models:

- **Uncertain and inaccurate information forms the basis for process reengineering**

  Information about current process structures is crucial in order to enhance process quality effectively. The more detailed, exact, and certain this information is, the better the modelling results will be. The problem lies in the conflict of aims between the (already mentioned) need for accurate information and the difficulties obtaining it, due to the often antiquated documents

---

[14] (Deiters and Gruhn and Striemer, 1995, p.460)

[15] See (Heilmann, 1994), (Karagiannis, 1994), and (Deiters and Striemer, 1995).

[16] (Huch and Schimmelpfeng, 1994, p.10)

[17] Such a cycle concept is described in (Deiters and Striemer, 1995), (Heilmann, 1994), (Karagiannis, 1994), (Raufer and Morschheuser and Enders, 1995), and (Scheer and Nüttgens and Zimmermann, 1995).

describing flow structure, varying or even contradictory statements from employees, and time constraints.[18]

Through the application of WFMS with an integrated protocol component, this lack of information could theoretically be filled. However, many contemporary systems do not have such functionality at all or not the necessary detail level. In addition there is no standard which defines what should be protocolled from the process instances and how it should be stored.[19]

- **Repeated process definition using WFMS**

Installing a WFMS and then putting it into operation is a costly configuration process. All flow definitions to be supported, must be defined detailedly within the WFMS. This results in a double workflow definition: The processes, modelled and simulated with a reengineering tool, must be modelled in a very similar manner once again with the WFMS.[20]

- **Long periods of adjustment for environmental changes**

Nowadays the environment of an enterprise can no longer be viewed as static. On the contrary, environmental changes are increasing. Thus the organisational structure cannot remain static, but must adjust to changing circumstances. Ideally these adaptation steps are constant.[21] However, in reality this is impossible partly for reasons concerning cost. It is more common to initiate organisational measures when greater environmental changes occur. The problem is, how can one identify this point in time? Matzenbacher has discussed the lack of objective economical instruments for its specification already in the 1970s.[22] Until now the situation has changed very little and in consequence many firms operate with a sub optimal organisation structure.

2.3 Workflow Management Cycle

This paper presents a new concept for the integration of modelling, execution, and monitoring of workflow based processes, the Workflow Management Cycle. It is the author's attempt at solving the problems already mentioned. This process management method does not rely on a specific research prototype, with which the problems concerning integration and evaluation could certainly be solved more easily. Instead the intention is to integrate commercially available tools into this

---

[18] (Krallmann, 1996, p.38)
[19] (Jablonski, 1995, p.20)
[20] (Jablonski, 1994, p.16)
[21] (Schertler, 1995, p.74)
[22] (Matzenbacher, 1979, p.275)

cycle model to demonstrate its practical use. This control loop model realises an effective link between modelling and execution of business processes. The cycle model comprises two constituents for different problems:

- A data integration component for reengineering tools and WFMS allows the almost completely automatic transfer of process models.
- The evaluation framework for process models allows a targeted examination of the workflow system's run-time-information. In this way disturbances to the normal control flow can be recognised early on.

The two main constituents are described in the next sections. A brief description of the whole cycle model follows.

2.3.1 Strategic and Operative Process Modelling

Process modelling can be divided into strategic and tactical/operative design tasks.[23] Generally speaking strategic modelling is the basis of for all further actions and is carried out by the corporate or division management. A competitive strategy which is fixed in advance and critical success factors derived from this form the basis for selecting and designing the core processes. These draft process models and other constraints are used in the following tactical process design.

Reengineering tools are well-suited for supporting strategic modelling. They allow an abstract modelling of the enterprise's flow structure. The ARIS-Toolset for example offers the possibility of creating Porter's value chain diagrams.[24]

Detailed planning of the flow structure takes place on tactical/operative level. Processes are refined to the level of individual jobs and the conceptual flow structure is depicted onto the enterprise's IT structure. Various experts are working together on this task. The economical process design is carried out largely by employees of the organisation department or external consultants. Depicting these process definitions onto the corporation's IT-structure is the task of specialised system administrators and programmers.[25]

Different tool classes are utilised for this task. Reengineering tools are useful for process design, while concrete workflow modelling, including resource allocation and design of data structures, tends to be 'carried out directly by the modelling front-end of the WFMS. This situation leads to a medium-break, which is eliminated by the WorkFlow Dictionary component as part of the cycle model.

---

[23] (Heilmann, 1994, p.14)
[24] The method handbook, volume 5, from the ARIS-Toolset documentation, describes this notation (IDS Prof. Scheer GmbH, 1995).
[25] (Galler and Scheer, 1995, p.23)

2.3.2 Process Evaluation

Regular evaluations of the organisation's current state are a fundamental characteristic of a feedback control circuit. Workflow systems help by automatically aggregating much temporal and quantitative data. The information collected is objective, precise, and actual and therefore very useful when taking planning steps. However, before the large amount of raw data can be used, it must be condensed, before an effective interpretation is possible.

Operating figures have often been used to compress large quantities of data. Adequate selected ratios show the essential aspects in a condensed and orderly manner and are therefore well-suited when making judgements and decisions.[26] The effectiveness of a ratio as a judgement aid depends greatly on the task to be supported. In terms of process evaluations this implies the necessity to identify the recipients of the information, so that different information, i.e. ratios, and their representation form can be selected.[27]

Within the framework of the Workflow Management Cycle, process data are used for the strategic as well as for the tactical/operative modelling. On the tactical/operative level, information with a short time horizon and lower aggregation is requested. The systems analyst requires information about single process types and instances. With the help of this information it is possible to enhance the detailed process definitions, e.g. since the analyst has more realistic time values for simulation runs.

Information showing intermediate term tendencies, i.e. the frequency of specific process types, is important for strategic planning. Focus lies on the flow structure as a whole and the intersections between different process types. This information was added to data from other control systems, such as information regarding cost and revenue accounting, and data from the enterprise's environment, e.g. general market and competitor information, to give an overall picture of the firm and its environment. Then indications for useful modifications to the competitive strategy can be derived.

The processing of raw process information is carried out by the WorkFlow Analyser. It implements a ratio model which is still being developed. *Executive information systems* (EIS) or *management information systems* (MIS) are often used as a presentation layer for this kind of information. In this context the EXECUdesk system is implemented in the author's department. This is a support tool for strategic planning of inter-company processes and is based on a distributed application architecture. The program is adapted to the needs of the corporate management.[28]

---

[26] (Lippold and Puhlmann, 1988, p.26)
[27] (Deiters and Striemer, 1995, p.12)
[28] (Müller-Wünsch, 1995)

Fig. 2.2 shows the cycle model described previously.

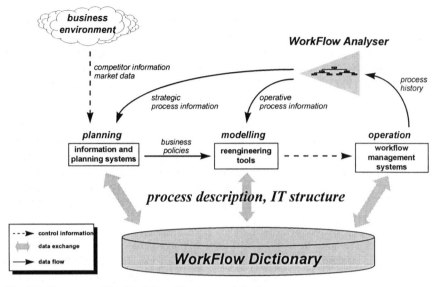

Fig. 2.2: Structure of the Workflow Management Cycle

3 WorkFlow Dictionary - Integration of Reengineering Tools and Workflow Systems

The WorkFlow Dictionary (WFD) is one of two basic components which make up the Workflow Management Cycle. Its purpose is the transfer of process models created during the modelling phase to the execution systems, i.e. the WFMS. The WFD is based on the idea of an universal integration model geared to the needs of many tools. In order to obtain or input data only a connection with the WFD has to be implemented for each tool. The WFD can be compared to a repository, a centralised meta data storage connecting the components of CASE-tools with one another.[29]

3.1 Integration Approach of the Workflow Management Coalition

The idea of integrating modelling tools and workflow systems was initiated some years ago by Jablonski.[30] Until now firms in particular have been the protagonists

---

[29] Compare for example (Habermann and Leymann, 1993).
[30] (Jablonski, 1994)

in developing interfaces between these tool classes. The problem is that all interfaces simply implement a 1:1-coupling of two specific tools.

In order to advance the standardisation in the field of WFMS as a whole and particularly also the interface between modelling tools and workflow systems, the open business consortium *Workflow Management Coalition* (WfMC) was founded.[31] One WfMC's work group is also developing an integration model, but this model only meets minimal requirements and does not satisfy practical demands.[32] The integration model is based on a modelling language which includes elements for its dynamically extension. Some firms which also participated in the development of this model, already use the extension elements to develop their own autonomous model. Demand for these model enlargements arise from the insufficient core model. In this way the idea of a common exchange model is only partially realised.

3.2 Structure of the WorkFlow Dictionary

The WorkFlow Dictionary is a meta model for storing process models. Within the WFD all aspects of an enterprise's organisational structure and flow structure can be stored, as well as specific concepts of workflow systems, such as roles and software applications. The WorkFlow Dictionary extends the approach of the WfMC in many ways. Beside common requirements[33] aspects of several other meta models were integrated, in particular those of the communication-structure-analysis[34] and the ARIS method.[35] When creating the model, the aim was to integrate all functions offered by various tools to the greatest possible extent, in order to support their full functionality. On the other hand great care was taken to avoid redundancy. It should not be possible to store equivalent fact in different ways.

An extended entity relationship notation is used in the model development. Nearly all relationships between entities have a n:m-cardinality, since it is the responsibility of the connected tools - and not of the WFD - to restrict the relationship.

---

[31] (Eckert, 1995) describes the objectives of this group and its current work status.

[32] (WfMC, 1995a) specifies the integration model detailedly.

[33] For example in (WfMC, 1994, p.5).

[34] In German: Kommunikationsstrukturanalyse (KSA).

[35] The ARIS-method is described in (Scheer, 1992). The KSA-method is the theoretical foundation of the BPR tool Bonapart and is described in (Krallmann, 1996, p.241).

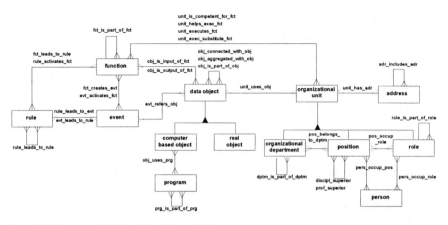

Fig. 3.1: Meta model of the WorkFlow Dictionary

The model is divided into three parts. The first section contains the entities and relationships which represent the organisational structure. The WFD extends the minimal requirements of a WFMS for example by offering ways to store the organisational hierarchy, places of organisational units and competencies. If possible, this information should be transferred into the target workflow system, since it is particularly useful for ad-hoc-modifications.

The second section allows the depiction of flow structures. In this section the model follows semantics of the ARIS-method, particularly the *extended event-triggered process chain*[36], a Petri-net-like notation.[37] The explicit representation of rules and events in this formalism allows the precise formulation of further conditions within a process.

The information view and resource view of processes make up the third model section. The entity *data object* is the central element of this section. Data is the processing object of all process tasks. Data is divided into *computer-stored objects* and *real objects*, i.e. not computer-stored objects. This distinction is necessary, because nowadays it cannot be assumed that all information is already stored electronically. The model must be able to maintain external references to real objects. Computer-based information can be assigned to programs, through which it can be processed, i.e. created, edited, printed etc.

---

[36] In German: erweiterte Ereignisgesteuerte Prozeßkette (eEPK).

[37] (IDS Prof. Scheer GmbH, 1995, volume 5: method handbook, chapter 4.4.) describes the eEPK-notation.

3.3 Prototype of the WorkFlow Dictionary

In order to verify the theoretical concepts, a prototype of the WorkFlow Dictionary is at present in the implementation phase. The relational database Gupta SQL Base[38] is used as the central storage facility for all model data. A connection with the WFD is made for the reengineering tools ARIS-Toolset V3.0 and Bonapart V1.45 and also for the workflow systems Hermes V2.0 and WorkParty V2.0.[39] Figure 4 outlines the WFD prototype.

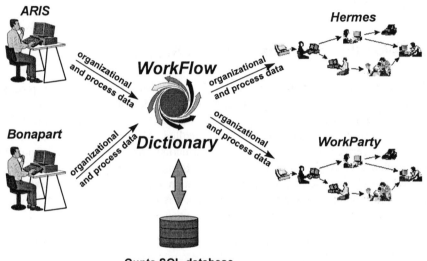

Gupta SQL-database

Fig. 3.2: Prototypical implementation of the WorkFlow Dictionary

4 WorkFlow Analyser - Evaluation of Business Processes

Many authors emphasise process evaluation as a crucial necessity for constructing a unified process management cycle[40], but hardly any running systems, or even methodical solutions are presented. Organisational theory has developed some

---

[38] Gupta SQL Base is a product from the Gupta Corporation, Menlo Park, USA.

[39] Hermes is a product from Carano GmbH, Berlin, Germany. WorkParty is a product from Siemens Nixdorf Informationssysteme AG, Paderborn, Germany.

[40] See for example (Karagiannis, 1994, p.112), (Deiters and Striemer, 1995, p.5), and (Scheer and Nüttgens and Zimmermann, 1995, p.431).

ratio-based models for the assessment of organisation structures[41], but these methods concentrate mainly on evaluating the company's organisational structure and ignore or at least neglect the flow structure. This deficiency is even more surprising, since although efficient process structures are widely demanded no measures exist for their evaluation.

At present no single method exists which meets the specific demands of workflow-based processes. Existing contributions in the field of IT-controlling are difficult to adapt since they concentrate on the demands of host-based data processing centres.[42]

4.1 Workflow Systems as a Data Supply

Workflow systems can provide a detailed protocol of all executed process instances and can therefore generate an exact model of the enterprise's flow structure. Only direct measurable values, such as time information, quantities, resource utilisation, or data use can be protcolled. All other indirect measurable values, i.e. derived values such as cost information, must be calculated from the protocolled data. Aggregation of this basic information is not a principal problem for WFMS. However, at present no uniform layout of the contents, extent of detail, and structure of such a process history is determined. A WfMC's work group deals with this topic[43], but so far no results have been officially published.

Besides the technical problems there are - in Germany at least - legal aspects to consider. The German *Betriebsverfassungsgesetz* (BetrVG) gives the works council the right to refuse installation of technical facilities, which enable performance of employees to be measured. The following section does not consider these problems.

4.2 Evaluation Framework of the WorkFlow Analyser

A pure outcome control is not sufficient for the effective monitoring of business processes.[44] Only intra-processual measurements lead to a detailed deficiency protocol. The intra-processual view is particularly important for the modeller, who needs in-depth information about each workflow. As regards corporate

---

[41] For example (Matzenbacher, 1979), (Lippold and Puhlmann, 1988), (Heilmann, 1990), (Saxer, 1993), and (Raufer and Morschheuser and Enders, 1995).

[42] Compare for example (Kargl, 1996).

[43] (WfMC, 1995b)

[44] (Gaitanides and Scholz and Vrohlings and Raster, 1994, p.58)

management however there is a demand for inter-processual information instead. This requires an overview of the flow structure as a whole.

The process evaluation framework is based on Picot/Reichwald's approach and is structured on three levels.[45] The first level or view, focuses on a single place of work. This allows examination of the elementary tasks within each process. The work place level has the lowest granularity degree within the framework. The process view on the other hand offers an insight into the process as a whole, whereby it is possible for example to compare single process instances. Lastly, the inter-process view presents information about the corporation's flow structure. Interdependencies between and redundancies among processes can be analysed. The process modeller needs all three views for his task, but the work place level and the process level are particularly important for optimising single processes. Corporate management relies mainly on the inter-process view, but however in terms of detailed consideration of drill-downs, a view of single process instances or even tasks within a process is also necessary.

To ensure a holistic view within each level the analyst has the choice of different evaluation dimensions:

- **Time view**

  The time view provides the typical measures of workflow, such as processing times, lay-times, or the total throughput time. Transportation times are less important on account of workflow systems. This information can be derived directly from the process history simply by totalling the time values from all work places.

- **Resource view**

  The resource view offers information about access and the access mode of data used in processes, i.e. read, write, create etc. This section also gives information about the use of programs and IT-infrastructure for process execution. Transmission costs of spatial distributed processes for example are vital for optimisation purposes.

- **Quantity view**

  The frequency of process execution is shown in the quantity view. Time values recorded by WFMS are accurate to the second, so that it is possible to analyse time variations during one day and likewise quarterly averages of process frequencies.

  Certain workflow-specific ratios are also encapsulated in the quantity view. Some values, such as the quota of early terminations, or the amount of ad-hoc-

---

[45] (Reichwald, 1987)

modifications of process types are extremely useful when judging the quality or adequacy of a process definition. The type of ad-hoc-modification can give hints regarding process reorganisation. Nastansky and Hilpert developed a classification scheme for this.[46]

- **Cost view**

Information regarding the approximate cost can be obtained by assessing work times and resource usage. This derived data is not as accurate as the outcomes of an activity-based cost assessment, but is completely adequate for rough calculations.

- **Personnel view**

The personnel view contains information about quantity and about the kind of elementary tasks within a workflow. Activity profiles of single employees can be compiled by using a preceding classification of these tasks, e.g. in planning, control, and work tasks, as is supported by some reengineering tools. These activity profiles and other measures, e.g. the number of different workflow types handled by an employee, can be used as an indicator of employee satisfaction.

The WorkFlow Analyser's ratio model is still being developed. Following completion of the model the intention is to validate the model against some real workflows. Fig. 4.1 outlines the procedure of process evaluation within the framework of the WorkFlow Analyser.

---

[46] (Nastansky and Hilpert, 1994)

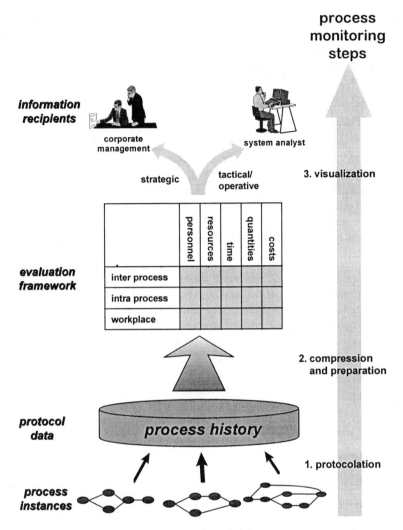

Fig. 4.1: Process monitoring with the WorkFlow Analyser

4.3 Prototype of the WorkFlow Analyser

Based on the evaluation framework a prototype of the WorkFlow Analyser will be implemented. The workflow system WorkParty serves as a data resource. It incorporates a protocol component, which provides all necessary information. The raw data, as well as the derived ratios are stored in the Gupta SQL Base-database.

The visualisation part will be implemented with pcEXPRESS[47], an EIS-shell. With this presentation layer the analyst can browse through all evaluation levels and dimensions. It is also possible to calculate daily, weekly, monthly, or other average values or viewing minimum or maximum throughput times of workflows.

5 Conclusion

With regard to the tendency towards process orientation, focus has until now been on modelling and designing the corporation's flow structure. However, effective process orientation requires support for the whole life-cycle of a process. This includes methodical and technical support, not only during process modelling but also in the implementation phase, for steering and controlling in operational work and for process monitoring. Workflow systems are an effective aid for the steering of process execution.

Transformation of process definitions from the modelling level to the execution level is supported by the WfMC's integration model and the more extensive WorkFlow Dictionary model. Methodical deficiencies constrain the evaluation of process structures, or organisational structures in general. The assessment of process quality and economic efficiency is a particularly significant problem. Restricting data accumulation to quantitative data and data which is easy to collect is a promising approach. In this way workflow systems are of great help in that they accumulate the raw data of the process instances. This basic information can be aggregated using operating figures.

References

Davenport, T. H. (1993) Process Innovation. Reengineering Work through Information Technology. Harvard Business School Press, Boston.

Deiters, W. and Gruhn, V. and Striemer, R. (1995) Der FUNSOFT-Ansatz zum integrierten Geschäftsprozeßmanagement. Wirtschaftsinformatik, 5/1995, vol.37. pp.459-466.

Deiters, W. and Striemer, R. (1995) Prozeßmanagementsysteme - Basistechnologie für ein entscheidungsorientiertes Informationsmanagement. Series ISST-Berichte 23/95. Fraunhofer Institute for Software and System Technology, Berlin, Dortmund.

Dier, M. and Lautenbacher, S. (1994) Groupware. Technologien für die lernende Organisation. Rahmen, Konzepte, Fallstudien. Computerwoche Verlag, Munich.

Eckert, H. (1995) Die Workflow Management Coalition. office management, 6/1995, vol.43. S.26-32.

Frese, E. (1992) Organisationstheorie. Historische Entwicklung - Ansätze - Perspektiven. 2nd edition. Gabler Verlag, Wiesbaden.

---

[47] pcEXPRESS is a product from IRI Software, Waltham, USA.

Gaitanides, M. (1983) Prozeßorganisation. Entwicklung, Ansätze und Programme prozeßorientierter Organisationsgestaltung. Franz Vahlen Verlag, Munich.

Gaitanides, M. and Scholz, R. and Vrohlings, A. and Raster, M. (1994) Prozeßmanagement. Konzepte, Umsetzungen und Erfahrungen des Reengineering. Carl Hanser Verlag, Munich, Vienna.

Galler, J. and Scheer, A.-W. (1995) Workflow-Projekte: Vom Geschäftsprozeßmodell zur unternehmensspezifischen Workflow-Anwendung. Information Management, 1/1995, vol.10. pp.20-27.

Götzer, K. (1995) Workflow. Unternehmenserfolg durch effizientere Arbeitsabläufe. Technik, Einsatz, Fallstudien. Computerwoche Verlag, Munich.

Hammer, M. and Champy, J. (1993) Reengineering the Coorporation. Harper Collins Publishers, New York.

Habermann, H.-J. and Leymann, F. (1993) Repository. Eine Einführung. Series Handbuch der Informatik - Volume 8.1. Oldenbourg Verlag, Munich, Vienna.

Hasenkamp, U. and Kirn, S. and Syring, M. (eds) (1994) CSCW - Computer Supported Cooperative Work. Informationssysteme für dezentralisierte Unternehmensstrukturen. Addison-Wesley Publishing Company, Bonn, Munich, Reading/Massachusetts.

Hasenkamp, U. and Syring, M. (1994) CSCW (Computer Supported Cooperative Work) in Organisationen - Grundlagen und Probleme, in CSCW - Computer Supported Cooperative Work. Informationssysteme für dezentralisierte Unternehmensstrukturen (ed. U. Hasenkamp and S. Kirn and M. Syring), Addison-Wesley Publishing Company, Bonn, Munich, Reading/Massachusetts. pp.15-37.

Heilmann, H. (1990) Ein Kennzahlensystem für die Organisation, in GI - 20. Jahrestagung. Volume 2. Series Informatik-Fachberichte - Volume 258 (ed A. Reuter), Springer Verlag, Berlin, Heidelberg, New York. pp.47-56.

Heilmann, H. (1994) Workflow Management. Integration von Organisation und Informationsverarbeitung. HMD - Theorie und Praxis der Wirtschaftsinformatik, 176/1994, vol.31. pp.8-21.

Huch, B. and Schimmelpfeng, K. (1994) Controlling: Konzepte, Aufgaben und Instrumente, in Informationssysteme für das Controlling. Konzepte, Methoden und Instrumente zur Gestaltung von Controlling-Informationssystemen (ed J. Biethahn and B. Huch), SpringerVerlag, Berlin, Heidelberg, New York. pp.1-24.

IDS Prof. Scheer GmbH (ed) (1995) ARIS-Toolset Handbücher. Version 3.0. IDS Prof. Scheer GmbH, Saarbrücken.

Jablonski, S. (1994) Ein integrierender Ansatz für Modellierung und Betrieb von Informationssystemen. Informationssystem-Architekturen - Rundbrief des GI-Fachausschusses 5.2, 1/1994, vol.1. pp.16-18.

Jablonski, S. (1995) Workflow-Management-Systeme: Motivation, Modellierung, Architektur. Informatik Spektrum, 1/1995, vol.18. pp.13-24.

Karagiannis, D. (1994) Die Rolle von Workflow Management beim Re-Engineering von Geschäftsprozessen. DV-Management, 3/1994, vol.4. pp.109-115.

Kargl, H. (1996) Controlling im DV-Bereich. 3rd edition. Oldenbourg Verlag, Munich, Vienna.

Kieser, A. and Kubicek, H. (1992) Organisation. 3rd edition. Walter de Gruyter Verlag, Berlin, New York.

Krallmann, H. (1996) Systemanalyse im Unternehmen. Geschäftsprozeßoptimierung, Partizipative Vorgehensmodelle, Objektorientierte Analyse. 2nd edition. Oldenbourg Verlag, Munich, Vienna.

Lippold, H. and Puhlmann, M. (1988) Zielsicher analysieren und steuern mit Büro-Kennzahlen. FBO Verlag, Baden-Baden.

Matzenbacher, H.-J. (1979) Zur Konzeption eines anwendungsorientierten Kennzahlenmodells zur Überwachung und Steuerung der Organisation. Zeitschrift für Organisation, 5/1979, vol.48. pp.275-281.

Müller-Wünsch, M. (1995) Der Management-Leitstand für das virtuelle Unternehmen: EXECUdesk. DV-Management, 4/1995, vol.5. pp.169-175.

Nastansky, L. and Hilpert, W. (1994) The GroupFlow System: A Scalable Approach to Workflow Management between Cooperation and Automation, in Innovationen bei Rechen- und Kommunikationssystemen (ed B. Wolfinger), Springer Verlag, Berlin, Heidelberg, New York. pp.473-479.

Österle, H. (1995) Business Engineering. Prozeß- und Systementwicklung. Volume 1: Entwurfstechniken. Springer Verlag, Berlin, Heidelberg, New York.

Porter, M. (1992a) Wettbewerbsstrategie (Competitive Strategy). Methoden zur Analyse von Branchen und Konkurrenten. 7th edition. Campus Verlag, Frankfurt am Main, New York.

Porter, M. (1992b) Wettbewerbsvorteile (Competitive Advantage). Spitzenleistungen erreichen und behaupten. 3rd edition. Campus Verlag, Frankfurt am Main, New York.

Raufer, H. and Morschheuser, S. and Enders, W. (1995) Ein Werkzeug zur Analyse und Modellierung von Geschäftsprozessen als Voraussetzung für effizientes Workflow-Management. Wirtschaftsinformatik, 5/1995, vol.37. pp.467-479.

Reichwald, R. (1987) Ein mehrstufiger Bewertungsansatz zur Wirtschaftlichkeitsbeurteilung der Bürokommunikation, in Wirtschaftlichkeitsrechnungen im Bürobereich. Konzepte und Erfahrungen. Series Betriebliche Informations- und Kommunikationssysteme - Volume 9 (ed R. Hoyer and G. Kölzer), Erich Schmidt Verlag, Berlin. pp.23-33.

Saxer, R. (1993) Monitoring des Informationssystems - ein Instrument zur Organisationsanalyse. Doctoraral thesis. University of St. Gallen, St. Gallen.

Scheer, A.-W. (1992) Architektur integrierter Informationssysteme. Grundlagen der Unternehmensmodellierung. 2nd edition. Springer Verlag, Berlin, Heidelberg, New York.

Scheer, A.-W. and Nüttgens, M. and Zimmermann, V. (1995) Rahmenkonzept für ein integriertes Geschäftsprozeßmanagement. Wirtschaftsinformatik, 5/1995, vol.37. pp.426-434.

Schertler, W. (1995) Unternehmensorganisation. Lehrbuch der Organisation und strategischen Unternehmensführung. 6th edition. Oldenbourg Verlag, Munich, Vienna.

Seghezzi, H. (1996) Integriertes Qualitätsmanagement. Das St. Galler Konzept. Carl Hanser Verlag, Munich, Vienna.

WfMC (ed) (1994) Workflow Management Coalition brochure. Brussels.

WfMC/Work Group 1/B (ed) (1996) Interface 1: Process Definition Interchange. Document Number TC-1020, Draft 5.0. Brussels.

WfMC/Work Group 5 (ed) (1995) Interface 5: Audit Data Specification. Document Number TC-1015, Draft 1.3. Brussels.

Process Modelling and Execution in Workflow Management Systems by Event-Driven Versioning

Waldemar Wieczerzycki

The Franco-Polish School of New Information and Communication Technologies, ul. Mansfelda 4, 60-854 Poznań, Poland, wiecz@efp.poznan.pl

Abstract. In the paper, a flexible persistent environment for business process modelling and management is proposed. A special emphasis is put on the evolution of process description over a given time, on one hand, and on the concurrent execution of transactions which correspond to activities embedded in processes, on the other hand.

The basic idea of the proposed approach is to extensively use object versions which are always considered in particular contexts, called the configurations. The configuration is both a granule of database versioning, a unit of database consistency and a granule of database distribution. When a business process is modelled, configurations substantially simplify the representation of its historical versions and current alternative versions. When a business process is executed, configurations support the avoidance and the resolution of conflicts which arise between concurrently executed transactions assigned to process activities.

1 Introduction

Business process modelling and management systems have their origins in a number of different developments to automate the business process in enterprises, notably document image processing and integrated office systems (Aiello et al. 1984; Brierley 1993; Bullinger and Mayer 1993; Hendley 1992; Jonnes and Morrison 1993). These systems have paved the way for the emergence of the workflow market in a receptive business environment.

Most of problems faced by enterprises concern internal business procedures that are neither well defined nor particularly efficient (Hales 1993; Lavery 1992; Medina-Mora et al. 1993). Service to customers needs to be improved and difficulties in introducing changes within the enterprise should be avoided. Business process modelling system is a computer-based, potential solution to these problems. It is a system for managing a series of tasks (actions) defined for one or more procedures. The system ensures that the tasks are passed among the appropriate participants in the correct sequence and completed within set times (default actions being taken where necessary). Participants may be people (actors) or other systems. People normally interact through workstations while other systems may be either on the same computer as the system considered, or on another accessible over a communication network.

Some of the business process modelling applications involve imaging, document processing and routing. These tasks can be effectively automated using current workflow automation products (Sheth 1994; Kling 1991; Hennessy et al. 1989). Some other applications involve tasks that must be modelled in a client-server style, using a transaction mechanism. These can be supported by distributed databases which, however, need to be adapted in order to better model the business process specificity (Georgakopoulos 1994; Huhns and Singh 1994, Schuster et al. 1994). Finally, business process systems are pro-active systems that embed long-living processes. Process agents, i.e. members of the organisation structures of the enterprise, are in charge of executing processes and process steps. Business process systems have to associate appropriate process agents to process pro-actively. That is why they should be built over active databases, which support those requirements in a natural way (Bussler and Jablonski 1994).

To be used as kernels of business process applications, distributed active databases require extensions concerning both: a data model, transaction management algorithms and distribution techniques. One of the main requirements addressed to the data model is its flexibility and the high level of modelling expression. To be used efficiently in business process applications, the data model must provide all necessary concepts in order to represent not only complex structure of processes, but also the process evolution over a given time (Benford 1991, Smith et al. 1989).

The process evolution may be seen in different ways. First, it may be seen in terms of improving the process description, elementary task allocation and scheduling, resource utilisation, etc., in order to increase the quality of services offered and, as a consequence, the total income of the enterprise. The improved process description typically replaces the previous one, next it is validated and finally, depending on the results achieved, accepted or rejected. In case of unsatisfactory results, process roll-back must be done in order to restore its previous description. It means that old descriptions must be somehow kept in the system. Even in the case of satisfactory results of process validation, an older description may be useful in the future, due to the cyclic nature of process or returning requirements.

On the other hand, process evolution relates to the creative nature of processes. Most of them usually create and develop very complex objects which are the expected outputs (artifacts) of processes. The progress is reached step-by-step through the creation of improved versions of the object being developed.

The second, important extension required in the database which is used in business process applications concerns transaction management techniques. New mechanisms supporting the efficient cooperation of long-duration transactions working in both shared and exclusive environments must be provided. The goal is obvious: conflicts between transactions of this type should be avoided whenever possible, and the system overhead related to it should be negligible. If, however, conflicts occur, they must be resolved in such a way that transactions are not suspended or rolled-back, but that their execution may be immediately continued.

Finally, the third extension required concerns distribution techniques in the database supporting business processes. Whenever a business process is modelled or executed, all objects accessed should be as close to the responsible user as possible, thus avoiding network transmission and delays. An adequate distribution granule reflecting the specificity of business processes must be used. Also the strategy of data distribution must match this specificity.

There are many approaches to business process modelling proposed in the literature (Brierley 1993; Bussler and Jablonski 1994; Garg and Jazayeri 1996; Hales 1993, Hendley 1992; Jones and Morrison 1993; Lavery 1992, Medina-Mora et. al. 1993; Sheth 1994; Steinmueller 1992). There is an evident lack, however, of approaches showing how a database might be used to support business process applications. Briefly speaking, in the paper a new approach for business process modelling and management in workflow applications supported by a database system is proposed. It provides all the extensions mentioned in the previous paragraphs.

The approach assumes extensive use of versions of particular object subsets. Up till now, different versioning models for object-oriented databases were considered in the literature (Ahmed and Navathe 1991; Atwood 1986; Cellary and Jomier 1990; Chou and Kim 1996; Estublier and Casallas 1994; Hubel et al. 1992; Katz and Chang 1987; Katz et al. 1986, Zdonik 1986). Most of them are, however, oriented for *CAD* applications. None is directly addressed to workflow applications. The paper proposes an enriched database model, influenced by the models mentioned above, which reflects the specificity of workflow applications.

Another contribution concerns the management of transactions which correspond to particular units of business process decomposition. Again, many different transaction models have been proposed up till now (Elmagarmid 1992; Garza and Kim 1988, Gray 1978), but, unfortunately, they do not take into account specific requirements of workflow applications. The paper extends existing transaction models by mechanisms increasing the efficiency of transaction execution which are strictly related to the data model proposed.

Finally, the paper tries to prove that a database offering a particular versioning technique may improve and simplify both process modelling and execution. Because of its generality, the approach may be applied also to other *CSCW* (Computer Supported Cooperative Work) applications, e.g. addressed to computer supported manufacturing, management and design systems (Baecker 1993; Bowers et al. 1988; Greenberg 1991; Grudin 1994; Johansen 1988, Kyng 1991).

The paper is organised in the following way: in Section 2 basic concepts are given and a new process modelling technique is proposed; in Section 3 new management techniques that may be applied to transactions corresponding to elementary process tasks are described; in Section 4 a particular approach to data distribution is presented; finally, in Section 5, the implementation of the approach is briefly discussed and conclusions are given.

2 Data Model

2.1 Basic Concepts and Conventions

In our approach we use some basic concepts commonly used in business process modelling systems. As it is well-known, a behaviour of every enterprise or organisation may be described by a set of more or less repetitive processes. In the frame of these processes, the enterprise employees cooperate by performing various elementary actions according to their skills and position in the organisational hierarchy. Usually, they share among each other the same working tools and other resources of the enterprise. Some actions may be performed simultaneously by different employees, while other are performed sequentially, i.e. they are ordered.

Briefly speaking, a model of an enterprise is composed of two parts: a model of organisational environments and a model of processes, which are strictly related to each other and mutually dependent (Hawryszkiewicz 1994).

Environments model the support structures for groups of employees. Environments can be of a prolonged duration and support a variety of processes. They can also define the social context for cooperation. Environments can include other environments, as well as artifacts, roles and actors:

- Artifacts represent the information base of the enterprise. These can include files as well as artifacts such as reports, designs, and so on. The artifact can include data values as well as methods to operate on these values, constraints and rules about the values, and policy rules used in the decision making process or in selecting appropriate roles to perform particular actions.
- Roles are abstract entities that represent system decisions. For example, purchasing a part requires a purchase requisition to be made and approved. The requisition and approval decisions are modelled as two separate roles.
- An actor is assigned to each role to make the decision. Actors are often positions that can be derived using organisational rules. Roles are not permanently associated with positions but are dynamically assigned using organisational rules.

Each *business process* (or simply *process*) corresponds to a subset of actions performed by the enterprise to achieve one of the goals it has been created for. Each process is within an environment and models interactions between a subset of objects in this environment. The notion of the process is not very useful because it does not contain details describing how the process is executed. Thus, it must be decomposed into smaller parts. The decomposition may be hierarchical, providing different abstraction levels of process description. The main unit of decomposition is that of an activity.

Activities correspond to different stages of process execution. They are partially ordered according to ways the process is executed in real-life. In order to be initiated, some activities require particular artifacts as an input, which may be taken directly from corresponding environments, or produced as outputs by other activities. Moreover, some activities are triggered by so called *events*, which may be classified into external events (e.g. a client's request), internal events (an employee's decision), and time events (a change of the year). In general, activities belonging to the same process may form a directed graph in which cycles are allowed.

In our approach, activities are only units of process decomposition. In general, however, they may be further sub-divided into so called *tasks*, which correspond to elementary, well-defined actions in the scope of a given activity.

48

To show the contents and a structure of a process we use the following graphic convention: processes are represented by soft boxes, activities by rectangles and the artifact flow that influences a partial order between activities by arrows. If two activities are joined by a single arrow (cf. Fig. 1a) then the situation is clear enough: the first activity produces an artifact that is necessary to run the second activity. Other typical cases are illustrated later on in Fig. 1. A single activity may produce two different artifacts that are used as inputs to two other activities (cf. Fig. 1b). It may also produce a single artifact that is logically shared by two other activities (cf. Fig. 1c). Sometimes a single artifact may be accompanied by the decision as to which activity, from a group of activities, it must be delivered; this is depicted by the use of an exclusivity arc (cf. Fig. 1d). Finally, to start an activity more than one artifact must sometimes be delivered (cf. Fig. 1e), or, it is sufficient to deliver one of the artifacts produced by a group of activities (cf. Fig. 1f).

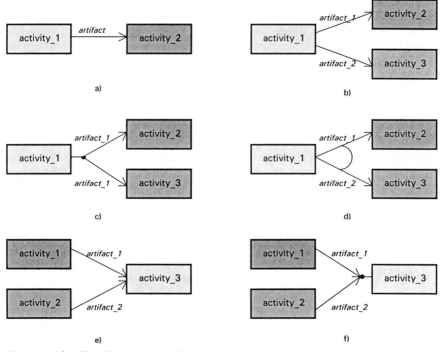

Fig. 1: Artifact Flow Between Activities

In general, activities belonging to the same process may form a directed graph in which cycles are allowed. As an example, consider a process of a document flow (cf. Fig. 2) which is composed of four activities: document creation assigned to a secretary role, document validation assigned to a head role, document correction assigned again to a secretary role, not necessarily the same as the previous one, and document sending assigned to a messenger role. Notice that two activities, document creation and document sending, are performed only once, while two other activities, document validation and correction, may be repeated many times depending on the result of recent validation.

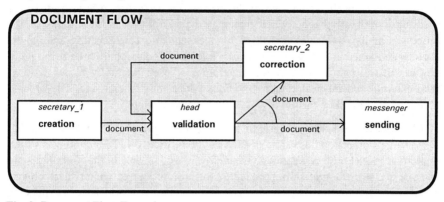

Fig. 2: Document Flow Example

In the next three sections we will show how the basic concepts presented in this section may be modelled in a multiversion database supporting business process modelling.

2.2 Process Initialisation

Every process is represented in the database by its precise description, which will be called the *process model*. This description contains the structure and behaviour of all objects that may potentially be used during process execution, or produced as a result of process execution, e.g. artifacts, roles. It also contains the exact specifications of all activities embedded in a process, their scheduling and activation rules, and their inputs and outputs. In database terms the descriptions of all processes modelled constitute the database schema.

At the logical level a database supporting business process modelling is a set of multiversion process models. A particular model, called the *root model*, which models an enterprise without processes is distinguished. It contains all environments which may be used to define a new process, which in turn are composed of objects like actors, resources available, standard roles that are well defined, however not performed yet, etc. In other words, the root model contains all objects which may be useful to execute future processes and can be foreseen since the beginning of enterprise existence.

Every process model describes exactly one enterprise process. Initially, i.e. after so called the *process initialisation*, it is composed of logical copies of some objects contained in the root model, and it is treated as a new version of the root model. The process initialisation is the first event that drives (triggers) automatic versioning in the database.

After initialisation, due to the detailed *process definition*, a process model is dynamically extended by new description objects local to the process being defined, reflecting the specificity of the process. In particular, it is extended by a specification of all the activities that may be performed in the frame of the process. Finally, it is also extended by the semantic relationships among the objects mentioned above, which order activities, assign them particular roles, point artifacts that are exchanged between activities or used internally by an activity, bind external and internal events to actions which they trigger, etc. The process definition stage will be further discussed in the next section (i.e. Section 2.3).

All processes are derived (initialised) from the same root model by selecting subsets of its objects. These subsets need not be disjointed i.e. some objects stored in the root model may be shared between processes. Updates in the root model are allowed and are automatically propagated to those processes which share objects affected. On the contrary, newly created objects in the root model are not included in the processes already initialised. They may be used in subsequent initialisations of new processes. Updates in processes are local, that means they do not influence the root model and models of other processes.

2.3 Process Definition

Every process model evolves during the database life-time in the same way corresponding execution rules are improved in the enterprise, in order to maximise its efficiency, profits, quality of services, etc. This process has a progressive nature, which means that it is performed step-by-step through the creation of improved versions of process models. The version derivation is driven by another type of event which is an explicit demand of the user who develops the process model.

New versions of process models do not replace older versions - they are kept parallel to them. Preserving old process model versions makes it possible to perform roll-backs in process model development, e.g. due to unsatisfactory results obtained. On the other hand, sometimes there is an evident need to return to previous versions of process models due to a requirements change or other factors which often have a cyclic nature, e.g. a process may be executed always in the same particular way on every December and differently during the rest of a year.

A database example in a way perceived at the logical level (i.e. by the user) is illustrated in Fig. 3. On the basis of the root model, four processes have been initialised. Next, the definition of every process has been refined by the creation of its new versions. Every process is represented in Fig. 3 by a stack of soft boxes; each box corresponds to a single process model version. Initial versions, filled in the same way, are put at the bottom of every stack. Also, the most recently released process model versions, which are put at the top of stacks, are especially distinguished. The models of process *P2* and *P4* share some objects; *P2* is available in three versions, while *P4* is available in four versions.

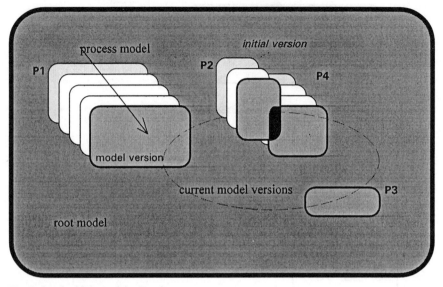

Fig. 3: Logical View of the Database

Because of different levels of abstraction, the logical view of the database is different than that of the database management system (physical view). In the latter case, one must discuss what is the granule of database versioning and what is the unit of the database consistency.

At the physical level the database is viewed as a tree of so called *database configurations* (cf. Fig. 4). Every configuration is composed of single versions of objects belonging to a subset of objects stored in the database, and corresponds to a single process model version. The root configuration corresponds to the root model. Every object is always addressed in the context of a previously chosen configuration. Updates performed on it are local to this configuration - they do not affect other versions of the same object belonging to other configurations. It also concerns updates performed on object versions that are physically shared between configurations; in this case a new object version is created which is local to the configuration addressed, while the old object version remains unchanged and is still available in other configurations. As a consequence, database configurations are mutually independent and they may evolve without imposing any constraints on the evolution of other configurations. The only exception concerns the root configuration; updates performed on shared objects belonging to it are propagated to respective child configurations.

The situation is different in the case of derivation of a new (child) configuration, as demanded by the user. The child configuration is initially

52

identical to its parent, i.e. it is a logical copy of its parent. Afterwards, it evolves independently without affecting objects in the parent configuration.

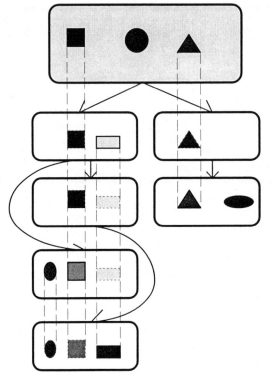

Fig. 4: Physical View of the Database

The afore-mentioned discussion follows that a set of database objects is the unit of versioning - every version of this set becomes a single database configuration. It is not possible to derive an object version out of scope of a particular configuration. Notice, that in two extreme cases a versionable set may be composed of all objects stored in the database, or a single object.

Finally, a database consistency problem has to be briefly discussed. The multiversion database is in general inconsistent, because as a whole it does not represent any valid state of the real-world being modelled. The notion of database configuration is the unit of database consistency, which means that object versions collected in the same configuration „go together". One may not say anything about the consistency between object versions occurring in different configurations.

2.4 Process Instantiation

One aspect is process modelling in order to precisely represent it in the database, and another aspect is process execution according to a previously accepted process model. In this section we will focus on the so called *process instantiation* which is the third event type driving automatic versioning. It starts the execution of activities embedded in the process according to patterns included in the process model. The emphasis will be put on additional modelling mechanisms rather than transaction management mechanisms, which we will discuss in Section 3.

As mentioned before, every process may be composed of an arbitrary number of elementary activities which are partially ordered. It is also assumed that, in general, processes are interactive and require intensive information exchange with users, who are modelled in the system as actors. One may distinguish three types of processes represented in the database. The first type contains read-only processes that read objects stored in the database, but do not update them and do not create new objects. Of course, processes of this type may produce new information, e.g. reports, letters, management directives, etc., displayed or printed to the users. The second type contains processes that may create new persistent database objects that are local to them, i.e. that exist only during process execution and play the role of artifacts exchanged between activities in the scope of the same process. The third, most general type, contains processes that may both update existing database objects and create new persistent objects that are retained in the database after process execution.

One of aims of the proposed approach is to avoid conflicts between processes when they are executed, and between processes and database operations described in the previous section, like process initialisation and definition. Most of conflicts arise when different processes try to access the same objects. Conflicts do not occur between processes of the type one case, they may occur between processes of the type two case, and they are very possible between processes of the type three case. The most straightforward solution to avoid conflicts is to isolate processes addressing them to different database configurations. It is relatively easy in the case of different processes, while not so obvious in the case of processes being instances of the same process model, which are executed asynchronously.

To avoid conflicts between instances of the same process model, a new database configuration is automatically derived whenever a process is instantiated, directly from the configuration which represents a corresponding process model. This new configuration becomes an exclusive scope in which a newly instantiated process is executed. Initially, it comprises logical copies of all objects included in the parent configuration, i.e. it contains all patterns included in the process model. Afterwards, it may evolve due to changes implied by process execution.

The process instantiation is illustrated in Fig. 5. Two processes, P_1 and P_2, are defined in the system. Process P_1 is represented by two model versions, while P_2

54

is represented by one model version. Process P_2 has three instances: *i1*, *i2* and *i3*, which are executed separately. Process instance *i1* is currently performing activity *a1*, instance *i2* is performing in parallel activities - *a2* and *a4*, while process instance *i3* has been finished. The last one has created three new objects, represented by circles, which are available only in the database configuration representing *i3*. Notice, also, that parallel to the execution of those three process instances, it is possible to refine the corresponding process model, which is included in the parent configuration logically independent from its children.

The following question arises: what to do with database configurations representing process instances after they are finished? In the case of processes of both the first and second type (as distinguished above), configurations may be simply removed from the system. In the case of a type three process, like *i3* in Fig. 5, one must decide what to do with objects created and modified by this process. In general there are two possibilities:
- the configuration becomes a new version of the process model, uniquely identified and visible to the users
- modifications made in the configuration are moved to the parent configuration, by merge or redo operations, and child configuration is deleted.

If the configurations become new versions then the number of versions of the same process model may increase very fast. Since there is definitely a difference between a configuration which represents a modified process model and a configuration which was created due to modifications of database objects via such a process, two levels of configuration addressing have to be available to the user. The first level concerns configurations corresponding to process model versions, while the second one concerns configurations created by instances of the same process, i.e. configurations sharing the same process model and varying only by objects being the deliverables of the corresponding process.

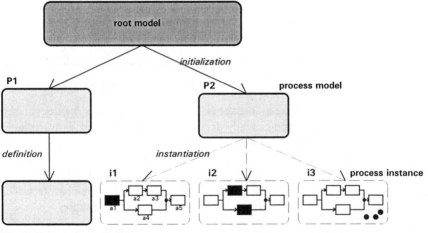

Fig. 5: Process Instantiation

To summarise, up till now three different events driving version derivation may be distinguished (cf. Fig. 5):

- process initialisation,
- user's request embedded in the process definition,
- process instantiation.

Process initialisation consists in the creation of a new database configuration on the basis of the root configuration. Process definition consists in refining a process model by adding to it local objects specific to the process and semantic relationships between them. It comprises operations on database objects intermingled with versioning requests which cause the derivation of a new child database configuration. Process instantiation first creates a new database configuration and exclusively dedicates it to a single process execution, and then starts the execution of respective activities embedded in the process.

3 Transaction Management

3.1 Transaction Model

A transaction is an elementary unit of interaction between the user and the *DBMS*. There are a lot of transaction models proposed in the literature. The only model used in practice, however, i.e. used in most commercial database systems, is the classical *ACID* model. As it is well known, in this model a transaction is considered as atomic and consistent, it works in isolation with other transactions and updates made during its execution are persistent. Ripe theory supporting the classical transaction model, as well as existing efficient transaction management algorithms, are encouraging enough to apply this model in our approach augmented, however, by some additional mechanisms. The same concerns classical concurrency control mechanisms which are based on lock setting (Gray 1978).

The main purpose of the mechanisms mentioned above, is to avoid, whenever possible, conflicts between transactions, and to avoid lock setting whenever it is not required, thus reducing the system overhead related with concurrency control. It is particularly important in the case of databases supporting business process modelling, in which one usually deals with so called long-duration transactions, which may sometimes last many hours or even days. Transactions of this type, in case of access conflict, may not be simply rolled-back or aborted by the *DBMS* because it would essentially be a waste of time.

In our approach we use a multiversion database, which, taken as a whole, is generally inconsistent. Thus, we extend the classical transaction definition in the following way: a transaction is a process that is addressed to a single database configuration and transforms it from one consistent state into another consistent state. We distinguish two following transaction types: model and activity transactions.

Model transactions are used at the initialisation and definition stages of a process life-time; they derive new versions of process models and/or update existing versions of process models.

Activity transactions are used during the process execution stage to perform operations described in a respective process model. An activity transaction corresponds exactly to a single activity of a process. It implies two important consequences. Firstly, we assume that an activity, similarly to a transaction, is atomic, i.e. may not be performed partially. This restriction may be relaxed if we allow an additional level of process decomposition, namely tasks, as explained in Section 2.1. Secondly, the execution of a process which is composed of n activities corresponds to the execution of n transactions in the same database configuration. Some of them are serialised in the same way corresponding activities are serialised, while others may be executed concurrently.

Now, one can briefly discuss conflicts that may occur between transactions of the two types mentioned above. Two model transactions conflict very rarely because they are usually addressed to different process model versions, i.e. they are addressed to different database configurations. If, however, they are addressed to the same configuration and they access overlapping subsets of objects included in the process model, then a conflict may occur. There are no conflicts between a model transaction and an activity transaction, even if they address the same process model version. In this case (cf. Section 2.4), the model transaction is executed in the database configuration that is a parent of the configuration in which the activity transaction is executed. Also, two activity transactions concerning different versions of a process model or different instances of the same process model never conflict. The only possible conflict between transactions of this type is when they perform two activities of the same process instance that may be executed in parallel (cf. instance *i2* in Fig. 5).

To summarise, conflicts between transactions may occur if they are physically addressed the same database configuration. In the next to subsections we show how conflicts like these may be avoided and how they may be resolved in case they occur.

3.2 Workspaces

Transactions addressed to the same database configuration tend to access subsets of objects, rather than all the objects stored in it. Moreover, objects included in these subsets are usually bound to each other by different semantic relationships, e.g. composition, inheritance. The subset of objects accessed by a single transaction is called its *workspace*. Workspaces accessed by different transactions in the same configuration may be disjointed or they may overlap.

Many of transactions performing workflow system activities access the same subsets of objects every time they are executed. Thus, *DBMS* has, a priori, the knowledge about transactions' workspaces. There are also transactions, usually highly interactive, whose workspaces depend on what the user requires from the workflow system. In this case, before transaction execution, the user may specify the workspace and give it to the *DBMS*.

From the user's point of view, the system should support two important functionalities. First, a flexible and powerful mechanism for defining a workspace has to be provided. Second, workspaces which overlap must be managed in a way that avoids, if possible, conflicts between different users accessing the same objects. A particular approach to workspace definition has been proposed (Wieczerzycki 95) in which the user defines his/her workspace iteratively, adding granules of the following types: a single object, an object with all its components, all instances of a single class, all instances of a single class together with their components, all instances of a class inheritance sub-graph, all instances of a class inheritance sub-graph together with their components.

If all the transactions that can be addressed to the same database configuration have their workspaces defined, then it is possible to determine whether a particular transaction requires locking during its execution or not. The decision depends on the following three factors:

- *Concurrency with other transactions.* If a transaction is executed concurrently with other transactions, like T2 or T4 in Fig. 6, then locks may be required and the next two factors must be considered. Otherwise, no locks are needed (cf. transactions: T1 and T5 in Fig. 6). This information may be obtained by calling the boolean system function parallel(Ti, Tj), which returns true if the transactions specified as parameters may be executed concurrently, and false in the opposite case.
- *Workspace intersection.* If two transactions have disjointed workspaces, then they never conflict and locks are not required. Otherwise, conflicts are possible and the last factor must be considered. This information may be obtained by calling the boolean system function conflict(Ti, Tj), which returns true if the transactions specified as parameters may access the same objects and at least one of them is not a query, and false in the opposite case.
- *Transaction state.* If all transactions that can be executed parallel to the transaction considered have already been committed, then there is no need for lock setting. This information may be obtained by calling the boolean system function committed(Ti), which returns true if the transaction specified as a parameter has been committed, and false in the opposite case.

To illustrate the above strategy consider transaction T_2 in Fig. 6. The lock setting must be applied to this transaction if the following expression evaluates to *true*:

(parallel (T2, T1) or parallel (T2, T3) or parallel (T2, T4) or parallel (T2, T5))

and

(conflict (T2, T3) and not committed (T3) or

conflict (T2, T4) and not committed(T4))

If a transaction requires the lock setting then some lock requests may be rejected due to their incompatibility with locks set by other transactions addressed to the same database configuration. In the next section we show how this situation may be resolved and transaction waiting or rolling-back avoided.

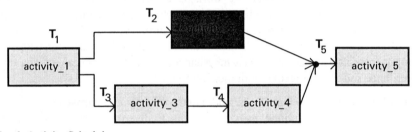

Fig. 6: Activity Schedule

3.3 Conflict Versioning

Assume two activity transactions which are executed in the same database configuration and potentially conflicting (i.e. the lock setting is necessary). What will happen if, after many operations on disjointed subsets of objects, those two transactions try to access the same object in an incompatible mode? In general, they are long-duration transactions, thus, aborting one of them is not recommended. The problem may be solved by providing a new database mechanism which is called conflict versioning.

Conflict versioning consists in the automatic derivation of a new private database configuration dedicated to a conflicting transaction (after detecting its first access conflict). The new private database configuration is a logical copy of its parent configuration which models process instance, with the exception that it contains all non-committed updates performed by the conflicting transaction, which are also logically removed from the parent configuration. At this point, the transaction for which the new configuration has been derived (i.e. the conflicting transaction) is not informed about this event and continues to access objects in the parent configuration. The system, however, automatically re-addresses all consecutive access requests to the newly created configuration. Now, locking conflicts in the private configuration are no longer possible. If, because of some reasons (e.g. user request), the transaction aborts, the private configuration is simply deleted. Otherwise, i.e. if the transaction commits, the user is informed by the system about configuration derivation, which becomes visible to other transactions.

After transaction commitment, the activity schedule graph must be analysed. If there is an activity (node) which requires artifacts from both separated subgraphs, then before starting it, the two database configurations must be merged. Otherwise, there is no need for merging and the process execution may be continued in two splitted configurations. The only exception to this rule is when the whole process execution is completed (i.e. when all process transactions are committed). In this case, in order to obtain a single consistent view of process artifacts (deliverables), the merge operation is necessary.

The conflict derivation technique for the process comprising five activities is illustrated in Fig. 7. Transactions T_2 and T_3 are executed concurrently in the same database configuration, directly after the commitment of transaction T_1 (cf. Fig. 7a). When access conflict arises, a new database configuration is automatically derived (cf. Fig. 7b): execution of T_2 is continued in the parent configuration, while execution of T_3 is continued in the child configuration. When T_3 commits, T_4 is initialised in the same configuration. Transaction T_5 requires artifacts from both subgraphs, which means that, after the commitment of T_2 and T_4, one must merge the two database configurations into a single one.

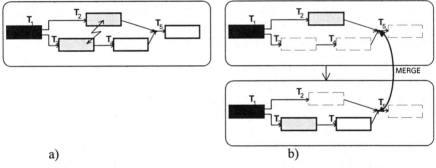

a)   b)

Fig. 7: Conflict Derivation

Two database configurations may be merged step-by-step by an object version comparison. This process, however, may be very tedious and usually requires many interactions with the user. Instead of this, a transaction **redo** may be performed. It consists in the automatic re-execution of operations performed in the child configuration once again in the parent database configuration, after committing all transaction addressed to the parent configuration. To support this mechanism, information about every transaction committed in the child configuration must be kept in a special file, usually called the *log-file*. During a *redo* operation, the contents of the log-file is interpreted and executed transaction by transaction in the parent configuration. After completing the last operation of the last transaction from the log-file, the *redo* process is finished and database configurations may be considered as merged. The *redo* operation is implemented as a single, multi-level transaction (Elmagarmid 1992). Sub-transactions nested in it correspond to transactions committed in the child configuration.

4 Distribution

4.1 Distribution Model

As it was mentioned (cf. Section 2.2), in our approach a single database configuration is a unit of both versioning and consistency. Thus, the most natural and straightforward approach to the object distribution is to treat a database configuration also as an elementary unit of distribution. This assumption may become very useful in the case of applications in which the user (or a group of cooperating users) tends to access a single database configuration or a set of database configurations, rather than all database configurations. Moreover, the user accesses most of objects stored in the database configuration addressed rather than a small subset only. This is a typical case of business process management applications in which every actor (user) is assigned to a particular process or a set of processes rather than to all processes modelled in the system. Moreover, when a particular process model is developed or a process is executed, the access to most of objects describing it is required, rather than to a small subset of objects.

In situations like those mentioned above, database configurations modelling a process developed or executed by a single user should be entirely stored on his (her) computer site, in order to avoid object version transmission via the network. As a consequence, when a particular object is addressed in a user's configuration, all other objects possibly bound to it by different semantic relationships are also immediately available, because they belong to the same database configuration. It is illustrated in Fig. 8: the horizontal axis represents multiversion objects stored in the database, while the vertical axis represent database configurations. Two configurations are stored on site *S1*, one on site *S2* and two on site *S3*.

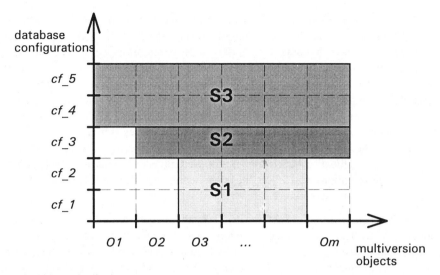

Fig. 8: Object Distribution

Before discussing the advantages of the proposed distribution model, we give some observations concerning relationships among database configurations stored on the same computer site, on one hand, and those stored on different sites, on the other hand.

Typically the number of database configurations in the multiversion database is much more greater than the number of computer sites, which means that in general on every computer site n configurations are stored. Moreover, database configurations stored on the same site are typically directly bound by the derivation relationship, because they are operated by the same user (a group of cooperating users). In other words, configurations stored on the same site typically belong to the same subtree of the derivation hierarchy. It is illustrated in Fig. 9. The correspondence between database configurations and computer sites given in Fig. 9a is much more typical and probable than that in Fig. 9b.

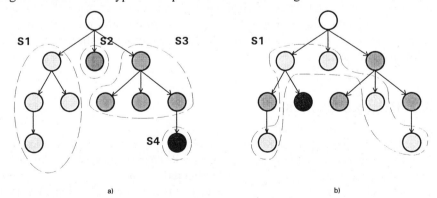

Fig. 9: Configurations and Computer Sites

Another observation concerns sharing of object versions between database configurations. Typically the highest level of sharing appears between configurations directly bound by the derivation relationship, i.e. parent and child configurations, due to the nature of the event-driven derivation. Transitively, there is usually also a relatively high level of object sharing between configurations indirectly bound by the derivation relationship, i.e. between configurations belonging to the same subtree of the derivation hierarchy. On the contrary, configurations included in different subtrees, e.g. subtrees stored on sites $S1$ and $S3$ in Fig. 9a, do not share object versions, or the sharing is very rare. As a consequence, in the distribution model proposed, object versions are mostly shared among configurations stored on the same computer site and sharing between different sites is incidental.

As it is known (cf. Section 2.3), database configurations are logically independent. Thus, an object version that appears simultaneously in two configurations stored on the same site is logically replicated, however physically there is no redundancy, and only one copy of shared object version is kept on the corresponding computer site. On the contrary, an object version that appears in two configurations stored on different computer sites is replicated both logically and physically. In other words, every site has its own physical copy of that object version. Those copies are independent, i.e. updating one of them does not impact the value of the second one. Due to the previous discussion, this replication does not substantially increase the level of redundancy in the distributed environment, providing that the assignment of configurations to computer sites corresponds to the one given in Fig. 9a.

The evident advantage of the proposed distribution model is that every time the user addresses his (her) database configuration, i.e. one of those stored on his (her) computer site, there is no need for network transmission, no matter if object is local or appears simultaneously in many configurations. In other words, every read, create or update operation may be performed immediately because a proper object version is always stored on the computer site the user is accessing. Notice, that even if the assignment of configurations to computer sites resembles the one presented in Fig. 9b, rather than in Fig. 9a, this feature is also preserved. In this case, however, the level of redundancy due to object version replication might be substantially higher.

4.2 Remote Access

Up till now we have assumed that the user addresses a database configuration that is assigned to his (her) computer site. What will happen, however, if, for some reasons, he (she) decides to access a configuration that is stored on a remote computer site? If the access request is incidental then an object version addressed may be transmitted over the network. Otherwise, i.e. if the user plans to access a particular remote configuration more regularly, e.g. due to remote process execution, the access request is treated by the system as an event of a new type driving versioning mechanisms, and the following steps are performed.

First a new database configuration is created from scratch which in the global derivation hierarchy becomes a next child of a parent node of the subtree assigned to the computer site accessed by the user (or a selected subtree, if more than one subtrees are assigned). Next, the remote configuration is entirely copied via the network to the one newly created, i.e. all object versions belonging to the remote configuration are physically replicated. This operation may cause some trouble due to object version transmission, proportionally to the number of objects existing in the remote configuration. To avoid it, so called the *lazy replication technique* is applied, which is particularly beneficial if the user accesses only a subset of objects from the remote configuration, and deletes the configuration at the end of his (her) session.

The lazy replication technique consists in keeping in the database a particular link between local and remote database configurations and delaying object version replication until they are explicitly addressed by the user (user's process). Because the remote configuration may not be frozen, i.e. it may continue to evolve independently from the local configuration, the link in fact is established between the local configuration and a *shadow configuration* of the remote one (both of them are stored on the same site) which is derived automatically. The shadow configuration is sort of consistent snapshot made at the time user requested an access to the remote configuration. The shadow configuration is stored in the system until last object version is replicated, or the local configuration linked to it is deleted by the user.

5 Conclusions

The main goal of this paper was to propose a flexible, persistent environment for business process modelling applications. A special emphasis has been put on the evolution of process description over a given time, on one hand, and on the concurrent execution of transactions which correspond to activities embedded in processes, on the other hand.

The basic idea of the proposed approach is to extensively use object versions which are always considered in particular contexts, called the configurations. The configuration is both a granule of database versioning, a unit of database consistency and a granule of distribution. When a business process is modelled, configurations substantially simplify the representation of its historical versions, which may be useful in the near future, and current alternative versions, which must be kept in the database simultaneously. When a business process is executed, configurations support the avoidance and the resolution of conflicts which arise between concurrently executed transactions assigned to process activities. Thus, the classical transaction model, common to most commercial database systems, may be preserved.

Configurations and special management techniques imply that the user need not be aware of object versioning. New database configurations are derived automatically by the system when particular events caused by users or transactions executed in the system occur.

The approach presented in this paper may be easily implemented over the Multiversion Object Manager (*MOM*) prototype developed at our institute, which is managed according to a particular versioning technique, originally proposed in (Cellary and Jomier 1990). The basic concept of this approach is that of a database version which comprises a single version of every object stored in the database. Every configuration may be modelled by a single database version extended by new concepts and mechanisms, as proposed in this paper.

References

Ahmed R. and Navathe S.B., *Version Management of Composite Objects in CAD Databases*, Proc. SIGMOD, pp. 218-227, Denver, Colorado, June 1991.

Aiello L., Nordi D., Panti M., *Modelling the Office Structure: A First Step Towards the Office Expert System*, Proc. of 2nd ACM SIGOA Conf., 1984.

Atwood T. M., *An object-oriented DBMS for design support applications*, Proc. COMPINT, pp. 299-307, Montréal, Canada, September 1986.

Baecker R., *Readings in Groupware and Computer Supported Cooperative Work*, Morgan Kaufmann, San Mateo, Calif., 1993.

Benford S., *Requirements of Activity Management*, in *Studies in CSCW: Theory, Practice and Design*, North-Holland, 1991.

Bowers J., Churcger J., Roberts T., *Structuring Computer-Mediated Communication in COSMOS*, in [eds.] Speth R., *EUTECO'88 - Research into Networks and Distributed Applications*, 1988.

Brierley E., *Workflow Today and Tomorrow*, Proc. of Conf. on Document Management, 1993.

Bullinger J.H., Mayer R., *Document Management in Office and Production*, Nachrichten für Dokumentation, Vol. 44, 1993.

Bussler C., Jablonski S., *Implementing Agent Coordination for Workflow Management Systems Using Active Database Systems*, Proc. 4th Int. Workshop on Research Issues in Data Engineering: Active Database Systems, 1994.

Cellary W., Jomier G., *Consistency of Versions in Object-Oriented Databases*, Proc. 16th VLDB Conf., Brisbane, Australia, 1990.

Chou H., Kim W., *A Unifying Framework for Version Control in CAD Environment*, 12 VLDB Conf., Kyoto, Aug. 1986.

Elmagarmid A. (ed.), Database Transaction Models, Morgan Kaufmann, 1992.

Estublier J. and Casallas R., *The Adele Configuration Manager*; In: *Configuration Management* , ed. by W. Tichy, Wiley and son, in Software Trend Serie, 1994.

Garg P.K., Jazayeri M, *Process-Centred Software Engineering Environments*, IEEE Press, 1996.

Garza J., Kim W., *Transaction Management in an Object-Oriented Database System*, Proc. ACM SIGMOD Conf., June 1988.

Georgakopoulos D., *Transactional Workflow Management in Distributed Object Computing Environments*, Proc. 10th int. Conf. on Data Engineering, 1994.

Gray J., *Notes on Database Operating Systems*, Operating Systems: An Advanced Course, Springer-Verlag, 1978.

Greenberg S., ed., *Computer Supported Cooperative Work and Groupware*, Academic Press Ltd, London, 1991.

Grudin J., *Computer Supported Cooperative Work: History and Focus*, Computer, May 1994.

Hales K., Workflow Management. An overview and some applications, Information Management and Technology, Vol. 26, 1993.

Hawryszkiewicz I., T., *A Generalised Semantic Model for CSCW Systems*, Proc. of 5th Int. Conf. on Database and Expert Systems, Greece, 1994.

Hendley T., *Workflow Management Software*, Information Management and Technology, Vol. 25, 1992.

Hennessy P., Benford S., Bowers J., Modelling Group Communication Structures: An Analysis of Four European Projects, In SICON 89: Proc. of the Singapore Conf. on Networks, IEEE Press, 1989.

Hubel C., Kafer W., Sutter B., *Controlling Cooperation Through Design-Object Specification, a Database-oriented Approach* Proc. of the European Design Automation Conf., Brussels, Belgium, 1992.

Huhns M., N., Singh M.,P., *Automating Workflows for Service Provisioning: Integrating AI and Database Technologies*, Proc. of 10th Conf. on Artificial Intelligence for Applications, 1994.

Johansen R., *Groupware: Computer Support for Business Teams*, The Free Press, New York, 1988.

Jones J.I., Morrison K.R., *Work Flow and Electronic Document Management*, Computers and Industrial Engineering, Vol. 25, 1993.

Katz R., Chang E., *Managing Change in a Computer Aided Design Database*, Proc. 13th VLDB Conf., 1987.

Katz R., Chang E., Bhateja R., *Version Modelling Concepts for CAD Databases*, Proc. ACM SIGMOD Conf., 1986.

Kling R., *Cooperation, Coordination, and Control in Computer Supported Cooperative Work.*, Comm. ACM, Vol 34, No 12, Dec. 1991.

Kyng M., *Designing for Cooperation: Cooperating in Design*, Comm. ACM, Vol. 34, No 12, Dec. 1991.

Lavery M., *A Survey of Workflow Management Software*, Proc. Conf. on OIS Document Management, 1992.

Medina-Mora R.,, Winograd T., Flores R., Flores F., *The ActionWorkflow Approach to Workflow Management Technology*, Information Society, Vol. 9, 1993.

Schuster H., Jablonski S., Kirsche T., Bussler C., *A Client/Server Architecture for Distributed Workflow Management Systems*, Proc. of 3-rd Int. Conf. on Parallel and Distributed Information Systems, 1994.

Sheth A., *Transactional Workflows: Research, Enabling Technologies and Applications*, Proc. 10th Int. Conf. on Data Engineering, 1994.

Smith H., Hennessy P., Lunt G., *The Activity Model Environment: An Object-Oriented Framework for Describing Organisational Communication*, Proc. ECSCW Conf., 1989.

Steinmueller B., *The JESSI-COMMON-FRAME Project - A Project Overview*, in Newman M., Rhyne T. (Eds): *Electronic Design Automation Frameworks*, North-Holland, 1992.

Wieczerzycki W., *Transaction Management in CSCW Applications*, to be published in the Journal of Systems Architecture.

Zdonik S.B. *Version Management in an Object-Oriented Database*, Int. Works. on Advanced Programming Environments, Norway 1986, pp. 139-200.

Scheduling Models for Workflow Management

Günter Schmidt

Information and Technology Management, University of Saarland, P.O.Box 151150, D-66041 Saarbrücken, Germany, gs@wi2.uni-sb.de

Abstract. Workflow management is the task to organise work on different activities in such a way that the underlying processes are carried out effectively and efficiently. Scheduling is the task of assigning scarce resources over time to competing activities such that given performance measures are optimised. Scheduling is an important part of workflow management in terms of optimisation but other features have to be added to scheduling systems to meet the requirements of workflow management. We will give an overview of these features, design a process model using a description language which is well known from project scheduling, and suggest some directions of how to model and solve scheduling problems within the area of workflow management.

Keywords. Workflow management, modelling, office scheduling

1 Introduction

Workflow management means planning, control, supervision, and execution of structured and predictable business processes in a distributed environment. These tasks can be supported by computer software called workflow systems. Modules for supporting scheduling decisions within such systems should be included to improve performance measures like meeting deadlines or due dates, minimising flowtime, or maximising capacity usage. Scheduling supports planning and control. It is the task of assigning scarce resources to precedence constrained activities over time best possible considering given performance measures. There is a need to merge the paradigms of either disciplines concerning the management of dependencies among activities (Malone and Crowston, 1994) to increase not only the potential of applying results from scheduling theory to the management of business processes but also to consider the relevance of special workflow problems within scheduling models. With this contribution we want to present a model which is suited not only to design workflow systems but also to formulate and solve the arising scheduling problems.

The basic data for scheduling are activities with precedence constraints, processing times, resource supply, resource demand, additional constraints, and various performance measures. The decisions to be taken are assignment and sequencing of activities considering resources and additional constraints. The same parameters and decisions are important for workflow management but more description is necessary to support the execution and supervision of processes by information systems. In addition descriptions for the specification of functions, data, required resources, organisational environment, and communicational needs are required. From the view point of workflow scheduling can be regarded as an approach to support the routing and sequencing of activities within workflow management. Scheduling in this environment we will also call office scheduling (Schmidt and Winckler, 1996) in order to separate the arising problems from manufacturing and computer processes.

In order to formulate workflow problems from the perspectives of information systems and scheduling we apply some basics of generalised activity networks (Elmaghraby, 1964). We will give some indication that this description language is powerful enough to help to design, optimise and implement business processes using current state of the art information technology.

The paper is organised as follows. We start with a framework for modelling information systems and define business processes and workflow. Then we develop a generic process model related to different views of an information system based on generalised activity networks incorporating scheduling aspects. Finally we use an example to demonstrate the modelling capabilities of our approach.

2 Modelling Information Systems

Modelling is a major task for analysis, design, and implementation of information systems. To be precise we state that the main part of an information system is its model because the model represents the capabilities of the system and determines its application. Building an information system means conversion of models. The analysis model is transformed via a design model into an implementation model which is finally transformed into a model of binary digits. The objective of an information system is to solve the type of problem which it is build for. In order to achieve this the system has to be modelled carefully answering the question

(i) which problem the system should solve and

(ii) how the system is going to solve the problem.

Either aspects behave like heads and tails of the same coin and such they should be investigated tightly coupled.

A framework for system modelling is given by the selected architecture. Quite generally an architecture represents the need to give a system a certain structure concerning its elements and their relationships. The selection of an architecture means to consider specific views on the system. Models of systems are build and then combined following the selected architecture.

An additional need of an architecture comes from model integration. Early approaches in this direction were developed in the field of computer integrated manufacturing. In the meantime new developments can be observed and a great number of architectures are proposed. Many of these ideas have been compared and evaluated with the objective to develop a generic enterprise reference architecture (Bernus and Nemes, 1996). Here the emphasis is not only to model information systems but also to support all activities connected to the lifecycle of an enterprise and its processes.

An architecture which is proposed to fit in the framework of such a generic enterprise reference architecture is LISA (Schmidt, 1996a). The major concern of LISA is to support modelling not only in the sense of formulating the problem but also in solving it. When using LISA these two aspects have to be represented explicitly and considered separately. They are connected by four different views on models for information systems. One view is concerned with the granularity of the system, one with the focused system elements and their relationships, one with the lifecycle of the system, and one with the purpose of modelling.

Concerning the granularity issue reference, enterprise, and application models are considered, concerning elements and relationships models for data, functional, and communicational aspects are developed, concerning the lifecycle of the system models for analysis, design, and implementation are used, and concerning the purpose we need models not only for the problem description but also for the problem solution. The problem description states constraints and objectives; the problem solution is a proposal how to meet them. The problem description is built on data, functions and possible communications. The problem solution is defined by the functional specification of steps to be carried out to convert some input into some output and by the realised communication on selected channels. Problem representation and problem solution can be modelled using different levels of granularity. Figure 1 gives an overview of the different views for information systems modelling based on LISA.

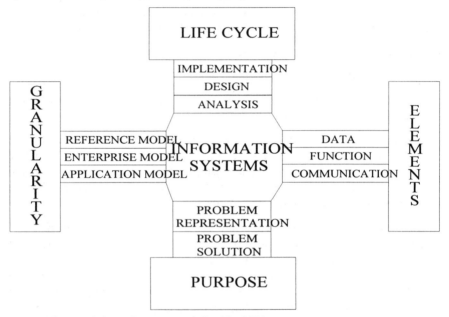

Fig. 1: Views on information systems defined by LISA

In order to build the different models an appropriate language is required. For the purpose of the analysis phase a language called GPN is proposed. Before this language is introduced in a later section we now discuss the relationship of business processes and workflow problems.

3 Business Processes

Workflow management and scheduling are both related to processes (compare the process paradigm of either disciplines in Blazewicz, Ecker, Pesch, Schmidt and Weglarz, 1996 and in Curtis, Kellner and Over, 1992, respectively). Processes are determined by activities for manipulating objects, their input and output, various kinds of resources needed for processing, and events starting and ending them. The way how processes are carried out is also determined by organisational constraints. These define which resources shall perform which activity applying which procedure. Resources for business processes are mainly humans, machinery (hardware), and information systems (software). Procedures are algorithms which are used by resources for processing activities. Questions to be answered about processes are what has to be done when, where, how, and why; who is doing it, and who needs the results.

An activity of a process may consist of a number of elementary tasks and transforms some input into some output. Activities or tasks are connected by precedence constraints which determine feasible sequences of execution. Starting and ending an activity is caused by events. Following a precedence relation requires communication between participating activities of a process in the sense that the output of a predecessor activity is part of the input of the successor activity.

Input and output of processes mainly exist as material (material processes) and information (information processes). For example the major shop floor activities in manufacturing have as input different kinds of raw material which are transformed to various types of output called processed material or products; office activities are mainly transforming data or information. In general input and output will consist of material and information simultaneously. Such hybrid processes are called business processes if they are market-oriented (Medina-Mora, Winograd and Flores, 1993). They can be defined quite generally like product development or customer order fulfilment; they also can refer to special lines of business like claims processing in insurance companies or loan processing in banks.

Each individual business process has to be designed in terms of activities, input and output, organisational aspects, data, communication, procedures, resources, and events taking into account business needs like quality, cost, time, new services and products. There exist many modelling methodologies for business processes which are either communication-based or activity-based.

Communication-based methodologies assume that the objective of business processes is customer satisfaction (Winograd and Flores, 1987). For each activity in a process the relationship between its consumer (sometimes also called customer, client, or principal) and its producer (sometimes also called performer, server, or agent) has to be modelled. Activity-based methodologies are more traditional and concentrate on the performance of an activity as the core modelling part. Most of the existing workflow models support the activity-based approach. The model presented here borrows elements from either approaches.

Once the processes are designed appropriately planning, control, supervision, and execution can be supported by information systems. Scheduling has to contribute to the planning and control function of such a system being connected to the supervision and execution of activities.

4 Workflow

In Georgakopoulos, Hornick and Sheth (1995) workflow is defined to be a collection of activities organised to accomplish some business process. An activity can be performed by one or more software systems, one or a team of humans, or a combination of these. This definition relates the workflow concept to automating the information part of business processes. There exist different kinds of workflow: ad hoc, administrative, and production workflows. Ad hoc workflows have no deterministic pattern for processing and sequencing activities. Administrative workflows consider repetitive, predictable processes; they are very often semi-automated where participants are actively prompted to perform their tasks. Production workflows have the same structure as administrative workflows but may be completely automated.

A workflow management system allocates work to participants Mahling, Woo, Blumenthal, Schlichter and Horstman, 1995). It coordinates the tasks carried out by the participant, determines the next participant, supplies necessary data and software and controls the workflow. They are tracking systems which make it possible for every participant to see status information about all activities and dependencies of a business process. Such systems were introduced as a commercial product in the early 1990s. They are further developed from electronic mail systems and active databases. Sometimes the focus is more on processes sometimes it is more on documents. In general the objective is the coordination of different business processes. Basic issues for workflow systems are the definition of activities and required resources, the capture of processing times and deadlines, the coordination of work results and the structure of the organisation.

5 Business Process and Workflow Models

There exist many activity-based approaches in describing business processes but none has been evaluated concerning its capabilities to support scheduling decisions (Schmidt, 1996a). We will suggest a modelling approach which has been successfully applied to support project scheduling. We will extend the modelling language such that parameters of a business process which are relevant for information processing can be represented. With this we can generate a model which is both capable of supporting the resource-oriented scheduling and the information-oriented execution aspects of workflow systems.

Our language to describe business processes is based on the concept of activity nets. An activity net is a graphical representation of activities including all relevant characteristics which are predominantly precedence constraints, time, and resource consumption. There exist the variants of (i) activity-on-node and (ii) activity-on-arc representation. In case of (i) all activities are represented by nodes and precedences between activities are represented by arcs; in case of (ii) activities are represented by arcs and nodes represent precedences and events starting or finishing some activity. Both representations result in a directed acyclic graph which is also called CPM model for deterministic durations of the activities or PERT model for stochastic ones.

CPM and PERT models assumes that (a) some activity can only be started if all predecessor activities are finished and (b) the structure concerning the relationship of activities is deterministic especially all activities must be performed. In case (a) is not true generalised precedence relations have to be applied (Elmaghraby and Kamburowski, 1992); in case (b) is not true so called Generalised Activity Nets (GAN) have to be used for modelling (Elmaghraby, 1964 and Elmaghraby, Baxter and Vouk, 1995). For the purpose of this paper we will discuss GAN in greater detail.

In GAN each node (event) has a receiving and an emitting part. The following receiving and emitting possibilities are defined:

AND: an event will occur when all activities leading into this node are processed;

IOR: an event will occur when at least one activity leading into this node is processed (inclusive-or);

XOR: an event will occur when one and not more than on activity leading into this node is processed (exclusive or);

MUF: all activities being triggered by this event must be processed (must follow);

MAF: an activity being triggered by this event may be processed taking into account a given probability (may follow).

With this we get six types of GAN nodes which are shown in Figure 2.

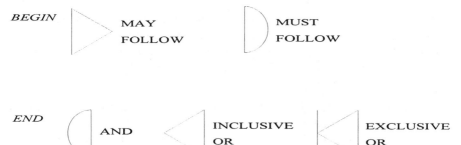

Fig. 2: Types of GAN nodes

The emphasis of this kind of activity net is to find answers on scheduling questions like timing, costing, and resource allocation within a process. Data, functional and organisational aspects of information processing are not treated in detail. Therefore we broaden the scope of these nets to describe business processes from an information processing perspective and call these Generalised Process

Nets (GPN). This is used as the basic modelling language for the analysis phase within the architecture LISA described in section two. A GPN consists of six layers and uses on the first layer the same structure of nodes and arcs as suggested within GAN. The second layer is dedicated to the functional specification, the third to communicational aspects, the fourth to data, the fifth to physical resources and products, and the sixth layer to the contract for performing an activity from an organisational point of view. The layer structure of GPN is shown in Figure 3.

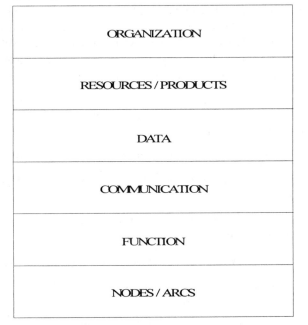

Fig. 3: Layers of GPN

Nodes and arcs relate to events and activities. We differ again between events starting activities and events finishing activities. They are labelled by
- the preconditions which must be fulfilled before starting and
- the postconditions which are fulfilled after finishing the activities,

Each activity is labelled by
- its name representing the function, procedure or algorithm to be executed in performing it and additional attributes including processing time, cost, etc.,
- the organisational unit which is responsible for performing the activity (producer),
- the organisational unit which is responsible for accepting the output of the activity (consumer),
- the physical resources required by the activity,
- the product generated by the activity,
- the data needed, and
- the data created.

Activities are being processed either by humans, by machinery, by software or by a combination of all of them. The different labels for events and activities are shown in Figure 4 using only MUF and AND nodes.

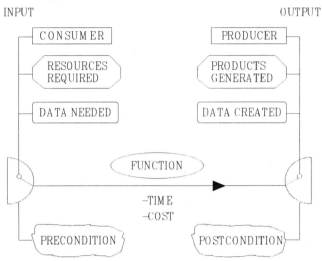

Fig. 4: Labelling events and activities for GPN

6 Procurement

We will now apply GPN to model a simple and linear office process which deals with purchasing goods and paying corresponding bills. If the manufacturing department (MD) is running out of stock for some item it is asking the purchasing department (PD) to order an appropriate amount of items. PD fills in some purchase order and transmits it to the vendor; a copy of this purchase order is given to the accounts payable department (APD). The vendor is sending the goods together with the receiving document to the ordering company; with separate mail the invoice is sent. Once the invoice arrives PD compares it with the purchase order and the receiving document. If the delivery is approved, APD will always pay the bill; if not PD starts complaining to the vendor. Invoices for purchased goods come in regularly and have to be processed appropriately. With each invoice very often discount chances and always penalty risks have to be taken into account. A discount applies if the bill is paid early enough and a penalty occurs if the payment is overdue.

This process can be speeded up using information technology. Applying concepts from electronic data interchange and from databases it is redesigned in the following sense. A purchase order is issued and stored in a database. From there it is transmitted by electronic data interchange to the vendor. On delivery of the goods the vendor transmits an invoice to the database of the customer where the purchasing order is stored; for efficiency reasons no receiving document is sent.

Following the process redesign the following issues have to be considered.

(1) An incoming invoice is classified by the following criteria: the amount due, the discount rate, the penalty rate, and the corresponding dates.

(2) In order to check completeness of an incoming invoice it is compared with the corresponding purchase order and the received goods. Each item of the invoice has to be verified.

(3) In order to check correctness of an incoming invoice it is checked concerning agreed prizes and additional purchasing conditions.

(4) If everything is correct the bill is ready to be paid in accordance with the liquidity constraints of the company; if not a complaining procedure is started.

The process is depicted in Figure 5 applying the notation of GPN and showing all five layers.

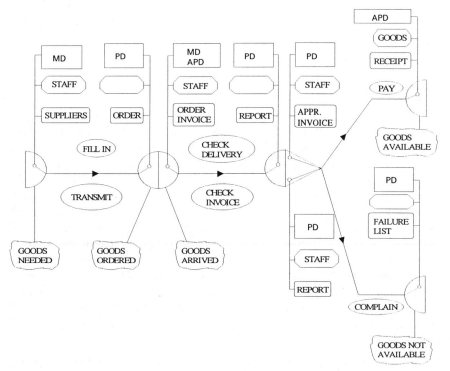

Fig. 5: Purchase process represented by GPN notation

7 Office Scheduling

There are many possibilities to optimise business processes. Two basic decisions influence efficiency and effectivity: one is the answer to the question how to design each process, the other is how to schedule competing instances of processes under the assumption of scarce resources. The above example corresponds to an office process (Ray, Palmer and Wohl, 1991) where the design decision already has been improved by applying electronic data interchange. Let us now concentrate on the scheduling decision.

In contrast to manufacturing environments office processes are not yet analysed in detail from a scheduling perspective (Blazewicz, Ecker, Schmidt and Weglarz, 1994). Quite generally office scheduling problems are concerned with the processing of documents. Documents can be considered as jobs to be assigned to resources such that given objectives and constraints can be fulfilled (Schmidt and Winckler, 1996). In terms of scheduling theory each job is related to an instance of a business process and consists of a set of precedence constrained tasks. Additional attributes to tasks and jobs can be assigned (Schmidt, 1996b). Typical performance measures for office processes are related to flow time, completion time and constraints are related to due dates or deadlines.

The language GPN can also be used to model business processes under the perspective of scheduling decisions. Let us consider a partial process of the above example related to purchasing of goods. It is the job we call bill processing consisting of two sequential tasks which are CHECK DELIVERY and CHECK INVOICE. Each task requires some processing time related to the work to be carried out. Moreover for each job two dates are known. One relates to the time for payment in order to receive some discount, the other to the time after which some additional penalty has to be paid. For the ease of the example we assume that discount and penalty rates are the same.

Let us further assume that there is only one resource i.e. one person of the staff available to perform these tasks and that there are three invoices waiting for processing. We will now show that the sequence of processing is of major influence on the payment considering discount and penalty possibilities. The following Table 1 summarises the problem showing invoice number (inv#), time to check (proc_time), discount date (disc_date), penalty date (pen_date), and the rate for discount and penalty (rate), respectively.

inv#	amount	proc_time	disc_date	pen_date	rate
1	200	5	10	20	5%
2	400	6	10	20	5%
3	400	5	10	15	5%

Table 1: Office scheduling example for bill processing

In general there exist n! possibilities to process n invoices using a single resource; for three invoices there exist six possibilities to sequence them on one resource. The range of the results for any of the six possibilities is from saving 30 units of cash discount up to paying additional 10 units of penalty depending on the sequence of processing. If there would be two resources available for this business process we had even more sequences and assignment possibilities with a corresponding range of results. Obviously, the result of the scheduling decision is of importance for the bill processing example. Using GPN we mark the partial process bill processing to indicate that it is subject to a scheduling decision (compare Figure 6).

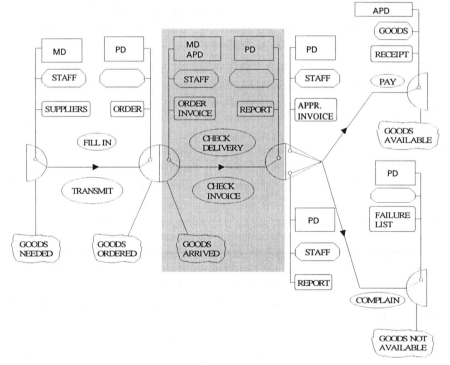

Fig. 6: Scheduling decisions within an office process

We have only considered a simple example of a business process which needs to be optimised from a scheduling point of view. Similar considerations can be carried out to analyse the effectiveness of various other business processes. There are a lot of them which have to be scheduled according to different criteria like time, cost, or quality. In general there are three scheduling decisions which have to be taken when a process is ready for execution:

(D1) sequencing for release to processing;

(D2) assignment to resources;

(D3) sequencing a particular resource queue.

In order to solve this kind of problems we refer to an extensive list on publications in the area of scheduling theory. For an introduction and an up to date survey of the main results we refer to Blazewicz, Ecker, Pesch, Schmidt and Weglarz (1996). For a presentation more related to the area of information systems we refer to Ecker, Gupta and Schmidt (1996) and Schmidt (1992).

8 Conclusions

We have given some indication that basic modelling languages from project scheduling can also be used to model selected business processes and corresponding information systems. We used general activity networks and generalised them to process networks with some added notation resulting in a six layer description. This representation can also be used for further optimisation purposes applying results from scheduling theory. One advantage of this approach is to combine process construction and process execution using only one notational framework and thus coming one step closer to the integration of models for problem representation and for problem solution. Additional work is necessary to evaluate this approach according various measures like completeness, consistency and ease of use.

References

Blazewicz,J., Ecker,K., Pesch,E., Schmidt,G., Weglarz,J. (1996), *Scheduling Computer and Manufacturing Processes*, Springer, Berlin et al.

Bernus,P., Nemes,L. (1996), Modelling and Methodologies for Enterprise Integration, Chapman and Hall, London.

Curtis,B., Kellner,M.I., Over,J. (1992), Process modeling, *Communications of the ACM* 35(9), 75-90.

Elmaghraby,S.E., Baxter,E.I., Vouk,M.A. (1995), An approach to the modeling and analysis of software production processes, *Int. Transactions Opl. Res.* 2(1), 117-135.

Ecker,K., Gupta,J.N., Schmidt,G. (1996), A framework for decision support systems for scheduling problems, *European Journal of Operational Research*, forthcoming.

Elmaghraby,S.E., Kamburowski,J. (1992), The analysis of activity networks under generalized precedence relations, *Management Science* 10, 1245-1263.

Elmaghraby,S.E. (1964), An algebra for the analysis of generalized activity networks, *Management Science.* 10, 494-514.

Georgakopoulos,D., Hornick,M., Sheth,A. (1995), An overview of workflow management: from process modeling to workflow automation infrastructure, *Distributed and Parallel Databases* 3, 119-153.

Malone,T.W., Crowston,K. (1994), The interdisciplinary study of coordination, *ACM Computing Surveys* 26(1), 87-119.

Medina-Mora,R., Winograd,T., Flores,R. (1993), Action workflow as the enterprise integration technology, Bulletin of the Technical *Committee on Data Engineering* 16(2).

Mahling,D., Woo,C., Blumenthal,R., Schlichter,H., Horstman,T. (1995), Workflow = OIS? A report of a workshop at the CSCW'94 conference, *SIGOIS Bulletin* 16(1), 59-64.

Ray,C., Palmer,J., Wohl,D. (1991), *Office Automation*, 2nd Edition, South Western-Publishing, Cincinnati, Ohio.

Schmidt,G. (1992), A decision support system for production scheduling, *Journal of Decision Systems* 1(2-3), 243-260.

Schmidt,G. (1996a), Informationsmanagement - Modelle, Methoden, Techniken, Springer, Berlin et al.

Schmidt,G. (1996b), Modelling production scheduling systems, *Int. J. Production Economics*, forthcoming.

Schmidt,G., Winckler,J. (1996), Office Scheduling, *unpublished manuscript*.

Winograd,T., Flores,R. (1987), *Understanding Computers and Cognition*, Addison-Wesley, Reading.

Introduction to Business Process Management Systems Concepts

Dimitris Karagiannis, Stefan Junginger and Robert Strobl

University of Vienna, The BPMS-Group, Bruenner Strasse 72, A-1210 Vienna, Austria, {dk, sjung, robert}@dke.univie.ac.at

Abstract. Business Process Management Systems (BPMS) are expected to meet the requirements of new designed business applications by
- supporting different *frameworks* which enable the modelling of current and anticipating future business application needs,
- integrating the existing and the new *information technology* enterprise environment, and
- providing a *continuous performance* method for assessment and improvement of the running business.

Their concepts, methods and technology (realised systems) should support developers who apply different frameworks (e.g. total quality management, re-engineering) to gathering and reasoning about various aspects in an organization in an efficient manner.

Furthermore this paper provides an overview of business process management issues and gives a quick description of a realised Business Process Management System named ADONIS.

1 Business Process Management Systems (BPMS) Methodology Framework

The BPMS-methodology is based on three abstraction levels; the *business level*, the *execution level*, and the *evaluation level*. These three levels correspond to the related graphs. The theoretical background and their interdependencies are described in chapter 2. The *operational model* of the BPMS-methodology is defined through a recursive *process-oriented model* which is based on the core activities *Criteria Selection, Information Acquisition, Analysis* and *Simulation, Design* and *Evaluation*. The BPMS methodology framework (Karagiannis 1989, Karagiannis 1994a, Karagiannis 1995, Kleinfeldt, Guiney, Miller and Barnes 1994) is tried and tested in different industrial projects. Experiences gained by some specialisations of the method provide the basis for the realization of specific application-oriented metamodels - especially in the banking and insurance area.

In Fig. 1 an instance of the operational model is given. According to that this BPMS-realization is viewed as a process consisting of five subprocesses, namely:
- Strategic Decision Process
- Reengineering Process
- Resource Allocation Process
- Workflow Management Process
- Performance Evaluation Process

The methodology framework mentioned uses all core activities, which are associated with each subprocess of the BPMS approach. Every such set of core activities should be individually refined, ordered and instantiated according to the requirements and the organizational aspects in the specific application field.

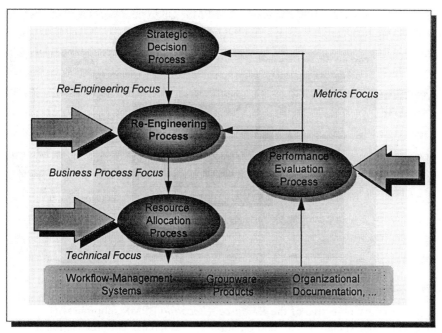

Fig. 1: BPMS-Approach: The process-oriented model

In the following a short introduction to the BPMS approach and in specific a description of the main subprocesses is given.

The *Strategic Decision* BPMS subprocess takes place after a strategic decision has been made for the reengineering of an enterprise's organisational environment. Based on global objectives, constraints for the processes to be selected are stated and success factors are recommended. The business processes are selected and the reengineering objectives are defined. Furthermore, the activities of initial information gathering and analysis concerning the selected business processes take place (Malone, Crowston, Lee and Pentland 1993).

The primary objective of the *Reengineering* BPMS subprocess (Fig. 2) is to design the new business process. Modelling constitutes a significant part of this BPMS subprocess, since the business process model to be generated has to be unambiguously defined, to further facilitate the execution of the following BPMS subprocesses. The designed business processes will have to conform with the evaluation criteria set in the *Strategic Decision* BPMS subprocess. Redesign takes place in an iterative way in order to obtain the best feasible results for the business process, keeping in mind all constraints relative to the business process, which might be imposed and affected by invariable factors. The *Reengineering* BPMS subprocess can be supported by a number of techniques, for instance modelling, simulation, animation, characteristic index calculation, etc. (Curtis, Kellner and Over 1992, Ellis and Wainer 1994). It can be further enriched by the actual use of new information technology as an enabler and facilitator (Yu and Mylopoulos

1994). In any case, the *Reengineering* BPMS subprocess has to cover human resource management issues, which might appear.

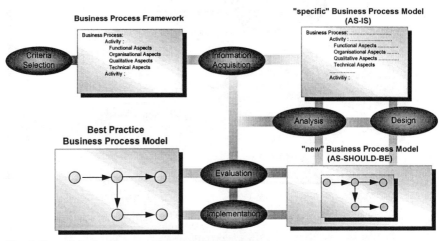

Fig. 2: Reengineering Process: The operational model

The primary objective of the *Resource Allocation* BPMS subprocess (Fig. 3) is to enable identification and coordination of resources, and realisation of the business processes (designed during the *Reengineering* BPMS subprocess). These resources are mainly related to Information Technologies, for instance, existing legacy applications might be modified, new applications might be implemented (Hewitt 1986, Casotto, Newton and Sangiovanni-Vincentelli 1990). All resource requirements should be readily derived from the results of the *Reengineering* BPMS subprocess.

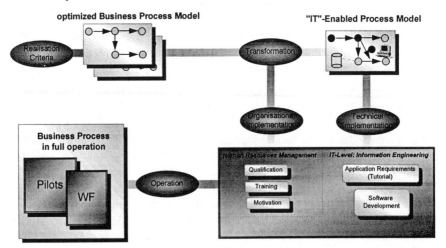

Fig. 3: Resource Allocation Process: The operational model

The primary objective of the *Workflow Management* BPMS subprocess is the execution of the reengineered business process, using resources made available during the *Resource Allocation* BPMS subprocess (Mahling, Craven and Croft 1995). New products, called Workflow Management Systems (Workflow Management Coalition Members 1994), have the requested abilities to define and control the workflow in an organization, transfer data, and integrate legacy information systems, existing software modules, and standard office software with the goal of facilitating a vendor's transformation to a „professional services" focus (Silver 1995). After test runs and additional corrective actions the process is executed in „real" time and location. This execution generates information necessary for the *Performance Evaluation* BPMS subprocess which follows.

The primary objective of the *Performance Evaluation* BPMS subprocess (Fig. 4) is the qualitative and quantitative evaluation of all information obtained by the realisation and execution of the reengineered business process. These results constitute an invaluable feedback for both the *Strategic Decision* and *Reengineering* BPMS subprocesses.

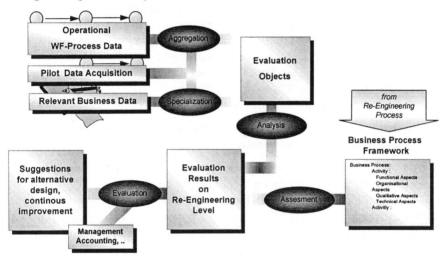

Fig. 4: Performance Evaluation Process: The operational model

2 Fundamentals of BPMS-Graphs

The BPMS-Paradigm, integrating the organizational, analytic as well as the IT aspects of business processes, is an approach for the management of business processes.

The main objective of this chapter is to point out how the different goals and tasks of BPMS-subprocesses in the course of modeling business processes influence the various models. It gives a definition of the contents of these models and provides a description how these models relate to each other with the main emphasis on the Evaluation Graph. In the following business process models are referred to as graphs.

Business process management starts with defining global strategic goals. During the *Strategic Decision Process* the business processes to be reengineered have to be selected.

In the *Reengineering Process* a business process model is designed according to the decisions taken at the former stage. The main objective of this stage (design task) is not a detailed modelling but to define a clearly structured and comprehensible design of the process including all important elements. The formal representation is called *Business Graph.*

During the *Resource Allocation Process* the available and necessary components have to be determined. As no particular modelling takes place in this process, organizational aspects are not to be regarded here. The information acquired will be used for modelling the business process, thus, the main emphasis is put on the execution of the business process in its real environment on the basis of its existing technical conditions. Therefore, a model is derived from the Business Graph (transformation task) which meets the operational and technical requirements. As this model is basically determined by the demands the execution has to fulfill its formal representation is called *Execution Graph.* In the following we assume that workflow technology is used due to the fact that the Execution Graph is always part of the workflow technology.

In the *Workflow Process* the business processes are performed according to the standards of their working environment.

In order to create a tool (Performance Evaluation Process) enabling us to set up evaluations and analytic statements about certain objects (evaluation task) on the basis of data registered during the performance of the Execution Graph (audit trail) a different view on the business process is necessary. This model contains all relevant information about the former graphs as well as their relation to each other shown by certain function graphs. The formal representation of this model is called *Evaluation Graph.*

Considering the different goals and therefore different views of each BPMS-process the modelling of business processes leads to different representations according to the particular situation. The relations of objectives, tasks, actors and the various BPMS-processes to the resulting graphs are illustrated in Fig. 5.

BPMS-Processes	Goals	Tasks	Actors	BPMS-Graphs
Strategic Decision	Fixing of the strategic business objectives	Product selection Market targets	Management Manager	
Reengineering	Design of the Business processes	Design Task	Management Organisation	*Business Graph* (organizational view))
Resource Allocation	Org./techn. implementation of the business process	Transform., Tutorial Implementation, org. measures (i.e. training)	Domains Data processing specialists	*Workflow Graph* *Execution Graph* (information techn. view)
Workflow	Product sale, Data acquisition	Execution	Domains Management organisation	
Performance Evaluation	Evaluation of Business processes Benchmarking	Evaluation Task	Management Management organisation	*Evaluation Graph* (analytically eval. view))
Strategic Decision ...	...	...	...	...

time

Fig. 5: BPMS-processes and -graphs: The relations

Organizational, strategic goals are in the foreground of the discussion. In the Strategic Decision Process objectives and visions of the enterprise have to be defined according to basic conditions. Criteria of success and quality must be determined.

Based on these guidelines three temporarily and logically consecutive stages of modelling can be specified. By means of the Reengineering Process the determined process model is the first to be developped, containing all data obtained at this stage and represented in the Business Graph.

At the second stage this model is - in accordance with the existing working environment and the available resources - transformed into a model coming up to these criteria in an efficient way. As it emphasizes the execution this model is called the Execution Graph.

The third stage provides the opportunity not only to calculate but also to evaluate certain objects of the above mentioned models. Therefore this model is called the Evaluation Graph.

The implementation of these graphs serves as the essential support to compare the assumed and estimated data with the results of the actual execution. It is now possible to adjust the given parameters to the real outcomes of execution with respect to the shown goals for the business processes. Having thus regarded the emergent principles from the execution the modelling process can be started again.

For better understanding we confined our investigations to the presented concepts and formalisms, which are based upon the BPR-tool ADONIS (Business Information Technologies Consulting GmbH 1996) (used for modelling the Business Graph) and upon the workflow-management-system IBM Flowmark for OS/2 (used for modelling the Execution Graph) in ADONIS respectively in IBM FlowMark (IBM Corporation 1994).

2.1 The Business Graph

In the Reengineering Process the determined process model[1] is built due to the results of the Strategic Decision Process. The determined process model is characterized by the following components: the temporary and logical subsequence of working units, the relevant objects of the working environment and the association between both parts of the model.

The business process modelling component contains the following objects:
- *Activities*
  Activities are the atomic units of a process, e.g. the working units which cannot or should not be divided any more.
- *Subprocesses*
  In order to achieve reusability and a higher level of abstraction - both necessary for distinctly structuring the process - activities are combined to subprocesses.
- *Control flow*
  The control flow - characterized by variables to which a particular probability distribution can be assigned - or also predicates determine the temporary and logical subsequence of the activities. Thus, parallelisms (or possibly their synchronization), decisions, sequences and loops can be described.

The following objects are part of the working environment model:
- *Actors (Persons)*
  This term represents the performers (not necessarily persons) of activities.
- *Groups*
  Groups are used to describe organizational structures such as positions, functions, roles, responsibilities and units of organizations.
- *Resources*
  Resources are defined as all means necessary for the realization of the activities such as documents, data or devices.

The mapping between business process and working environment can be specified as a set of relations, e.g. as the following:

---

[1] This model consists of two parts: Business Process Model and Working Environment Model.

- *„Person responsible for"*
  This term defines who is carrying out an activity.
- *„Is required"*
  This term regulates which resources are required for carrying out an activity.

The representation of the determined business process arising from this procedure is called Business Graph. The Business Graph can be formally described as the following:

Let $\Gamma = \{\beta_j, \omega, \rho_{ij}\}$ be a Business Graph.

We denote:

a number of business processes $\beta_i = \{A_i, \beta_j, \kappa_{ij}\}$

where $i \in \{1,2,3,....,n\}$ and $j \in \{1,2,3,...,m\}$ in which

A_i signifies activity i, $i \in \{1,..,n\}$

β_j signifies subprocess j, $j \in \{1,..,m\}$ and

k_{ij} a control connector ij, $i \in \{1,..,n\}$. $j \in \{1,..,m\}$

a working environment $\omega = \{P_i, G_j, R_k, \pi_{ij}\}_i$,

where $i \in \{1,2,....,n\}, j \in \{1,2,..,m\}$ and $k \in \{1,2,..,o\}$ in which

P_i signifies a person i, $i \in \{1,2,....,n\}$

G_j signifies a grouping j, $j \in \{1,2,..,m\}$

R_k signifies a resource k, $k \in \{1,2,..,o\}$ and

π_{ij} the relations between the elements of P_i, G_j, and R_k, , $i \in \{1,2,....,n\}$,

$j \in \{1,2,..,m\}$

and the relations ρ_{ij} between the elements of β_i und ω.

2.2 The Execution Graph

The main concern of the second stage is to transform the model corresponding to the actual possibilities and the existing environment, i.e. information technology applied to the process determines the modelling. As a rule activities have to be divided; but it is also possible to integrate several of them into a new one or to design their performance in a different way.

The normal procedure to build this new model is the following:

Starting with the Business Graph a semi-automatic transformation and a corresponding tutorial lead to the so-called Workflow Graph. Final sophistication and improvement complete the building of the new model which meets all data processing requirements. It is characterized as follows:

- *Activities*
  The semantic working units are called activities, but are not necessarily identical with those described in the Business Graph. Although showing a certain relation they have no bijective mapping.
- *Control structures*
  The performance of each activity is defined by the following attributes: The starting conditions indicate when an activity can be started, the transitory conditions regulate which activity has to be performed next and the ending conditions determine the end of an activity.
- *Subprocess/block*
  Together with their control structures several activities can be integrated either to a subprocess or a block. This procedure serves several purposes: first it guarantees an easier reusability of certain subprocesses and hence a better control, second it allows the modelling of loops (by building blocks with corresponding ending conditions).

As an essential difference as regards the first model the following objects are taken into consideration:

- *Programs*
  Software and applications needed for the execution are to be defined (data, localization) and to be assigned to particular activities.
- *Data format*
  regulates which kind of data has to be available and how the data flow has to be organized. The activities can be taken as representations of their data-input-container to their output-container as it is suggested by (Leymann and Altenhuber 1994).
- *Documents*
  This term includes storage mediums of all kinds such as written documents, pictures or sound carriers.

Corresponding to the working environment model of the Business Graph the following objects describing the organization have to be enumerated:

- *Organizational structures*
  This term includes objects such as organizational units, persons, functions, rolls, and competences.
- *Relations*
  Regulates the relations between the objects of an organization: Organizational units are characterized by hierarchy, persons by competences. Persons can act on behalf of another person, can be member or coordinator of a function, can be member or manager of an organizational unit.

This business process, which is enriched with IT technology aspects, builds the Execution Graph and can be formally described as follows:

Let $E = \{\varepsilon_i, o, v_{ij}\}$ be an Execution Graph.

We denote:

a number of processes $\varepsilon_i = \{ B_i, \varepsilon_j, \lambda_{ij} \}$,

    where $i \in \{1,2,....,n\}$ and $j \in \{1,2,...,m\}$ in which

    B_i signifies activity i, $i \in \{1,2,..,n\}$

    ε_j signifies subprocess j, $j \in \{1,2,..,m\}$

    λ_{ij} an attribute ij, $i \in \{1,2,..,n\}, j \in \{1,2,..,m\}$

a working environment $o = \{O_i, P_j, R_k, K_l, o_{ij}\}$,

    where $i \in \{1,2,....,n\}, j \in \{1,2,...,m\}, k \in \{1,2,...,o\}$ and $l \in \{1,2,...,p\}$

    in which

    O_i signifies an organizational unit i, $i \in \{1,2,....,n\}$

    P_j a person j, $j \in \{1,2,...,m\}$

    R_k a function k, $k \in \{1,2,...,o\}$

    K_l a responsability l, $l \in \{1,2,...,p\}$

    o_{ij} a relation ij, $i \in \{1,2,..,n\}, j \in \{1,2,..,m\}$

and the relations v_{ij} between the various elements.

This definition of an Execution Graph in a generic framework allows an independent description of a Business Process Execution. By using a Workflow Management System for processing this Execution Graph, e.g. FlowMark, a mapping between the generic Execution Graph and this specific one should be performed. This specific Business Process Execution Graph in FlowMark is presented as a straightened, weighted and coloured graph.

2.3 The Evaluation Graph

The third stage (the Performance Evaluation Process) aims at creating a tool enabling us to calculate and to evaluate interesting objects. The main objective is to obtain statements about the determined process model, as it has to be investigated and must be optimized by that time. In general we distinguish between two cases: for one thing the data result from the Business Graph itself (as for instance from the simulation or from the entries in the notebooks), for another - and this is possibly more important - the data can result from the performance of the process in the existing environment (registered objects such as time: activity A was started at 12 and terminated at 12,15). Regarding the first case the knowledge about the structure of the Business Graph is sufficient to perform the necessary analyses. As regards the second case however, special assignments between Business Graph and Execution Graph and their relevant objects have to be made in order to ensure that the calculations can be applied from one model to the other. In the following figure the „dynamic mapping" action indicates where evaluation should be realized.

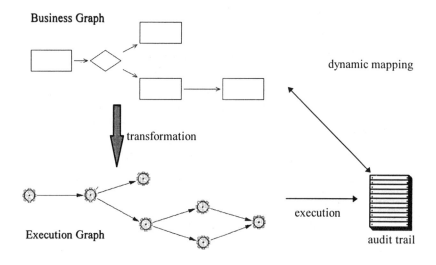

Fig. 6: Evaluation Task

The Evaluation Graph is considered an extension of the Business Graph, due to the fact that on the one hand it has to contain the sequence of the activities and their relations with the organization, on the other hand however it has to offer large-scale functionality[2].

A set of „assignment rules" ψ_i between the Business Graph Γ and the Execution Graph E build the main extension in comparison to the definition of these graphs. In the set theory this situation is defined as shown below.

The Evaluation Graph Θ can be defined as a superset of the Business Graph and respectively the Execution Graph, as Γ can be considered as an isomorphic graph to Γ and $\hat{E}$ as an isomorphic graph to E which will be shown in detail in part 2.5.

$$\Theta = \hat{\Gamma} \cup \{\psi_i\} \cong \hat{E} \cup \{\psi_i\}, i \in \{1,..,n\}$$

Summarized, in order to apply the requested improvement of business processes we have to define the Evaluation Graph Θ which results from the graph Γ in

---

[2] Mapping- and assignment rules.

connection with graph E under consideration of the „operational semantics" aspects about the business processes (Hinkelmann and Karagiannis 1992).

2.4 Inter-graphs Dependencies

Due to the temporary and logical creation of the various graphs the following structural interdependency between them arises as it is shown in Fig. 7.

Starting with the determined process model a corresponding transformation and a tutorial lead to a Workflow Graph. By means of different changes (e.g. sophistication and increase of information) - meeting the data processing requirements - the Execution Graph is obtained.

Exactly as the Workflow Graph is an intermediate stage during the transition to the Execution Graph, the Business Graph forms the basis of building the Evaluation Graph.

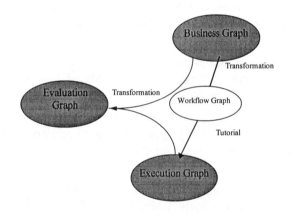

Fig. 7: Structural interdependencies between the graphs

2.5 Building the Evaluation Graph

The intent to use data resulting from the Business Graph (for instance time or costs) for analytical statements about the determined process model guarantees - because of the knowledge about the structure of the graph - that all requirements for the evaluation are met. Yet, in order to be able to relate the data resulting from the Execution Graph (for instance from the audit trails) to the corresponding activities in the Business Graph, representation rules are necessary which assign the appropriate activities and control structures of the Execution Graph to the activities and control flow of the Business Graph. These functions also describe which kind of interdependencies arise between the objects of both graphs that are to be evaluated.

In general the procedure for building the Evaluation Graph runs like follows:

To the given Business Graph Γ and Execution Graph E one has to identify a set of activities with their control structures of Γ with a set of activities with their control structures of E. These sets are called evaluation parts or briefly parts which meet the following preconditions:
1. A part is coherent, i.e. every activity of a part consisting of at least two activities has either a predecessor or a successor.
2. Parts are disjunctive tuples, i.e. one part consists of only one activity.
3. Joining the parts with the control connectors being derived from the initial graph results - apart from isomorphism - in the orginal graph, i.e. one part consists of only one activity.

By choosing the right parts these conditions can always be fulfilled as both initial graphs meet all the preconditions. In general a profound sophistication of these parts is indicated (in ideal circumstances one part consists of only one activity) in order to receive evaluation statements which are as substantial and accurate as possible.

As a result of integrating the various parts in their graphs into subprocesses and building control structures between them which were induced from the original graphs one gets from Γ to the isomorphic graph $\bar{\Gamma}$, respectively from E to the isomorphic graph $\hat{E}$.

The identification of the corresponding parts leads to a set of assignment rules ψ_i which determine a unique inverse mapping ψ between the Business Graph $\bar{\Gamma}$ and the Execution Graph $\hat{E}$.

In order to obtain such a mapping ψ, the following important and fundamental cases are possible in which A respectively A_i signify the activities of the Business Graph, B respectively B_i signify the activities of the Execution Graph. How this assignment rule determines the relations between the objects to be evaluated is given as an example for the operation time (T_i respectively t_i signify the operation time of activity A_i respectively B_i).

1. Activity A consists of several activities B_1, .., B_n ,which are to be carried out consecutively.

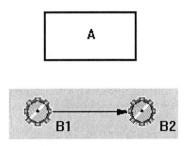

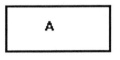

ψ: $\{A\}$ <-> $\{B_1, B_2; \lambda_{12}\}$

For the times holds true: $T_1 = t_1 + t_2$.

2. Activity A consists of several activies B_1, .., B_n , which are to be carried out simultaneously.

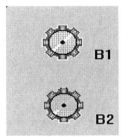

ψ: $\{A\}$ <-> $\{B_1, B_2; \lambda_{12}\}$

$T_1 = \max \{t_1, t_2\}$.

3. Activity A is described by a decision making process.

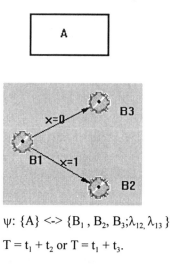

ψ: $\{A\}$ <-> $\{B_1, B_2, B_3; \lambda_{12,} \lambda_{13}\}$

$T = t_1 + t_2$ or $T = t_1 + t_3$.

4. The activities $A_1, ..., A_n$ are combined to activity B.

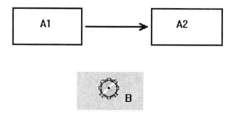

ψ: $\{A_1, A_2, \rho_{12}\}$ <-> $\{B\}$

$T_1 + T_2+ = t$. It is particularly evident (and a significant statement of the assignment function ψ), that the time T_1 by itself cannot be described by objects of the Execution Graph. This is a general rule for combining activities.

5. The activities A_1, ..., A_n being carried out simultaneously are considered in a single activity B.

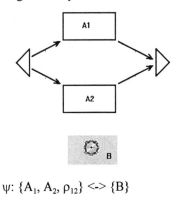

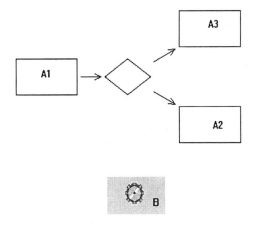

ψ: $\{A_1, A_2, \rho_{12}\}$ <-> $\{B\}$

max $\{T_1, T_2\} = t$.

6. A decision making process is represented by activity B.

ψ: $\{A_1, A_2, A_3, \rho_{12}, \rho_{13}\}$ <-> $\{B\}$

$T_1 + T_2 = t$ or $T_1 + T_3 = t$.

Parts which cannot be deduced from these fundamental cases can also be described by a mapping function. As a result however, the relations between the evaluation objects result in a higher grade of complexity.

It is a general rule that the evaluation objects coming from the part of the Business Graph must be equal to those coming from the corresponding part of the Execution Graph.

2.6 Generate the Evaluation Graph: An Example

The following example illustrates the procedure for constructing the Evaluation Graph. Given is a section of a Business Graph Γ - shown in Fig. 8 - as well as the corresponding Execution Graph E. The object to be evaluated is the operating time (as regards the modelling of the Business Graph), i.e. the sum of the operation times inside the path beginning with activity A_1 and ending with activity A_{10}. Given are also the recorded (Audit Trail) times $t_1, ..., t_9$.

As a first step the parts of the Business Graph are identified with the corresponding parts of the Execution Graph (indicated by double arrows in Fig. 8). By this, the necessary assignment rules ψ_i are known.

Combining the parts of the various graphs to subprocesses we get the graphs $\hat{\Gamma}$ and $\hat{E}$ as shown in Fig. 9. Those parts which consist of only one activity are not represented as subprocesses.

If S_i respectively s_i signify the operation times for the subprocesses in the Business respectively the Execution Graph, the following rule applies to the operation time T^*:

$$T^* = S_1 + S_2 + T_6 + \max \{S_3, T_7\} + T_{10},$$

or described by the help of the audit trail times

$$T^* = t_1 + t_2 + t_5 + \max \{t_6, t_7\} + s_2, \text{ in which } s_2 = t_8 + t_9.$$

Hence we are able to calculate certain objects as for instance the operation time in this example, by means of the objects of the audit trails. At the same time the representation function ψ informs us which objects cannot be calculated by the values coming from the Execution Graph, as for instance the time T_1 in our example.

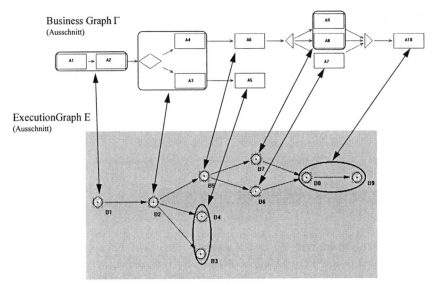

Fig. 8: "Semantical mapping" of E and Γ Graph

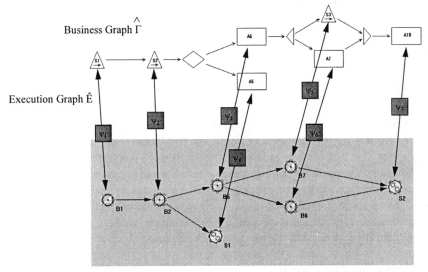

Fig. 9: The unique inverse mapping ψ in order to generate the Evaluation Graph

3 Building BPMS-Applications: The REFINE Project

The applicability of the BPMS-methodology and the functionality of the ADONIS toolkit are now to be presented by describing the REFINE-project (Reengineering Insurance Business), Project No. 20588, (REFINE-Proposal 1995), one of our international activities in the 4th European Union IT-Programme Framework under participation of Austria, Germany, Greece and Spain.

3.1 A Customized BPMS-Methodology

REFINE's main strategic goal is to enable the insurance sector to face the deregulated and liberalised European market. This will be accomplished by customising and refining the concepts of Business Process Reengineering for the insurance sector, using Information Technology (IT) as an enabler for achieving efficiency and flexibility.

The strategic goal of the REFINE project is to contribute significantly to the accomplishment of the main objective delineated in the workprogramme of the 4th EU Framework. More specifically, the project partners intend to use information technology as an enabler, in order to increase the competitiveness of European Insurance Enterprises within global markets through the application and use of best business practice in the areas of business process methodologies, tools and human resources.

The main purpose of the REFINE project is to effectively transform core insurance business processes by applying information technology in order to produce readily applicable, reusable results in the wider European tableau. The generic Business Process Management Systems (BPMS) methodology framework will be used as a guideline during the process of transformation of the selected business processes. This framework will be customized in order to serve the special needs of insurance enterprises. The specific outcome of this effort is the transformed specific business process. This output as well as generic outputs gained from the transformation experience can be reused by other organisations in the insurance and in the wider financial sector in order to improve their own operations and competitiveness.

The technical and business objectives of REFINE using BPMS and the ADONIS toolkit are to:
- demonstrate the effective transformation of selected core insurance business processes at the insurance sites,
- apply an advanced business methodology and use leadership information technology as an enabler for the realisation of the transformation of selected business processes,

- produce results applicable to every day business operations,
- produce results to be reused by other insurance enterprises in the wider European market in order to improve their operations and competitiveness.

In order to ensure the accomplishment of these objectives, the following factors constituted the foundation upon which the project has been composed:
- the selected business processes are common across all participating insurance companies and are of high interest to all European insurance enterprises,
- all insurance partners have the same objectives for the best practice pilots to be realised at each insurance site,
- a common business process methodology will be used by all partners, based on the BPMS methodology framework,
- result evaluation is based on metrics methods to be applied by the insurance companies.

Major development phases

The project is divided into four major development phases which are realised in different workpackages, performed by the project partners.

The main development phases of the project are:
- modelling and analysis of selected business processes,
- design and implementation of new business processes,
- consolidation and dissemination of results.

In the first phase the selected business processes at the participating insurance sites were modelled and analysed. Based on the analysis results the second phase is now concerned with the design of new processes. In this phase different solutions will be elaborated and evaluated using some refined evaluation metrics. The best solution is implemented during the third phase. This phase also covers the customization of the transformation plan.

The consolidation activities of the results from the participating insurance sites, where the reengineering pilots take place, is performed after every intermediate result of the project is obtained. A continuous dissemination of the project results ensures a wider applicability of the REFINE work and results. Thus, dissemination is an activity which, similar to consolidation, also takes place in parallel with the execution of the previous three phases, according to the exploitation and the dissemination plan of REFINE.

Intermediate results to be disseminated and by which the progress of the project can be evaluated are:
- Models of the selected business processes at the participating insurance sites,
- analysis reports of the modelled processes at the participating insurance sites,
- descriptions of the proposed business processes solutions,
- evaluation criteria for the proposed business processes solutions,
- specification of the optimum solution per user site,
- a process transformation plan per user site,

- documentation of implementation of the proposed business process solution (i.e. the reengineered business processes),
- reengineering evaluation metrics.

3.2 ADONIS-Toolkit: A BPMS Implementation

The ADONIS toolkit supports the core activities of the BPMS framework. The BPMS concepts in ADONIS are realized within a metamodel shown in Fig. 10 using C++. The Booch OO-Method is used for the toolkit conceptual design. The modelling within the toolkit is based on a business process- and a working environment metamodel (Kühn, Karagiannis and Junginger 1996). A business process language named ADL (ADONIS Definition Language) enables the interchange of designed business models and realized business libraries within the environments. Note that with ADONIS the business process model as well as the working environment can be designed either with a graphical component or by using the ADONIS Definition Language (ADL).

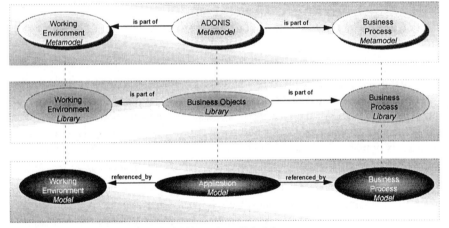

Fig. 10: The ADONIS-Metamodell: Hierarchy of Models

ADONIS is a multi-user client/server toolkit for modelling, analysis and evaluation of business processes and their underlying organizational structure. Its functionality can be briefly described as:

- Acquisition of business process information:
  Information Acquisition Component
- Representation of business processes and organizational structures:
  Modelling Component
- Complex queries about the modelled environment:
  Analysis Component
- Performance check and what-if analysis:
  Simulation Component

- Support of business process implementation using workflow management technology:
  Transformation Component
- Feedback analysis of realized workflow models:
  Evaluation Component
- Interchange of business models & libraries using the ADL-language:
  Import/Export Component
- Facilitating global business by supporting local languages (German, Greek, English, French, Hungarian, Italian, Portuguese, Spanish):
  Multi-Language Component

3.3 Experience

One of the major results of the project will be two reengineered process instances of the proposed business processes, realised at three different user sites. The selection of an optimum solution (i.e. a new process) is based on certain evaluation criteria, which are also part of the project result, since they are refined within the project. The implementation of the selected optimum solution takes place according to a transformation plan, also refined within the project, to satisfy the reengineering goals. The reengineering results are evaluated against the initial reengineering goals using refined evaluation metrics. Thus, the end results of REFINE are:

- the two reengineered processes at three different user sites,
- experiences gained during best practice pilots, which can be used for the development of an insurance-specific reengineering methodology,
- evaluation metrics for selecting an optimum solution (i.e. a new reengineered process) and metrics for evaluating the end result of process reengineering.

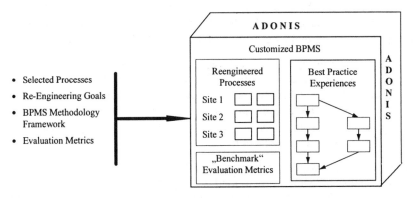

Fig. 11: REFINE: An insurance customized BPMS

In Fig. 11 the BPMS approach is shown in direct correspondence to a customized model that can be used to support the different subprocesses and the various elements of the methodology framework utilised as a basis for the development of process-oriented business applications.

Reengineering of selected processes from the insurance area is carried out according to the BPMS methodology framework (Karagiannis 1994) using the BPMS working environment which has been applied to a number of industry applications. It is shown how to use a methodology framework rather than a specific methodology, as there are no mature reengineering methodologies in the commerce specifically for selected sectors, as the insurance area.

4 Conclusion

The goal of this paper is not only to support researchers and/or IT-developers to clarify and integrate the different business process approaches in a BPMS-framework but also to help business-oriented people in practise to „ ..... develop a coherent theory of re-engineering - to look at what worked from the company point of view" as Gora (Garrett 1995) requests.

Major issues have raised that affect business process in a distributed environment in view of abstraction, modelling, and mapping and its impact on the realization of business process management systems.

The remaining question is whether an operational semantic can be defined between the BPMS-graphs that can be utilized to apply a „correct" interpretation of the results produced by a business environment.

In this paper we have not provided detailed solutions but have enumerated the issues that need to be considered in order to design a BPMS that supports the realisation and running of distributed business applications.

Our main effort in the future will be to present a first draft of a conceptual model about the three abstraction levels as well as a theoretical background for the definition of the corresponding graphs and for generating the Evaluation Graph.

Furthermore a rule base (Karagiannis and Hinkelmann 1994) should be developped which describes how certain objects in the evaluation can be related from one model to the other. In this connection the assignment rules ψ_i , we looked upon as given, are of great importance. In fact, the information emerging from the transition from the Business Graph to the Execution Graph is sufficient for the definition of these assignment rules.

However the following questions should be considered from the software engineering point of view: which kind of algorithms can be used for this identification and to what extent can this already be realized during the transformation from the Business Graph to the Workflow Graph respectively during the transformation to the Execution Graph.

Acknowledgements

The European Community supported parts of this work through the REFINE-project no. 20588. We are especially grateful to Rainer Telesko and the BPMS-Group at the University of Vienna for helping us clarifying this paper. We also thank Gabriele Kaiser for her help.

References

Business Information Technologies Consulting GmbH (1996) ADONIS für OS/2, Benutzerhandbuch.

Hewitt, C. (1986) Offices are Open Systems. *ACM Transactions on Office Information Systems*, Vol. 4, No. 3, 271-287.

Curtis, B., Kellner, M.I. and Over, J. (1992) Process Modelling. *Communications of the ACM*, Vol. 35, No. 9, 75-90.

Casotto, A., Newton, A.R. and Sangiovanni-Vincentelli, A. (1990) Design Management based on Design Traces, in: *Proc. 27th ACM/IEEE Design Automation Conference*, 136-141.

Ellis, C.A. and Wainer, J. (1994) Goal-based models of collaboration. *Collaborative Computing*, 1, 1-86.

Garrett, T.M. (1995) Re-engineering the Insurance Company: Two perspectives. *LOMA's Resource magazine*.

Hinkelmann, K. and Karagiannis, D. (1992) Context-Sensitive Office Tasks: A Generative Approach. *Decision Support Systems 8*, 255-267.

IBM Corporation (1994) IBM FlowMark for OS/2, User Manual.

Karagiannis, D. (1989) Flexible Bürosysteme (FBS): Architektur und Einsatzmöglichkeiten. *Technischer Bericht FAW-TR-89001, Forschungsinstitut für anwendungsorientierte Wissensverarbeitung*.

Karagiannis, D. (1994) Towards Business Process Management Systems. *Tutorial at the International Conference on Cooperative Information Systems (CoopIS '94), Toronto*.

Karagiannis, D. (1994a) Die Rolle von Workflow-Management beim Re-Engineering von Geschäftsprozessen. *dv Management 3/94*, 109-114.

Karagiannis, D. (1995) BPMS: Business Process Management Systems: Concepts, Methods and Technology. *SIGOIS Special Issue, SIGOIS Bulletin*, 10-13.

Kleinfeldt, S., Guiney, M., Miller, J.K. and Barnes, M. (1994) Design Methodology Management, in: *Proceedings of the IEEE*, Vol. 82, No. 2.

Karagiannis, D. and Hinkelmann, K. (1994) Sharing of very large-scale knowledge bases: A rule-selection approach, in: *Knowledge Building and Knowledge Sharing* (eds. K. Fuchi and T. Yokoi), IOS Press, Tokyo, 182-191.

Kühn, H., Karagiannis, D. and Junginger, S. (1996) Metamodellierung in dem BPMS-Analysewerkzeug ADONIS. *Business Information Technologies Consulting GmbH*, 1996.

Leymann, F. and Altenhuber, W. (1994) Managing Business Processes as an Information Ressource. *IBM Systems Journal*, Vol. 33, No. 2, 326-348.

Malone, T. W., Crowston, K., Lee, J. and Pentland, B. (1993) Tools for inventing organizations: Toward a handbook of organizational processes, in: *Proceedings of the*

2nd IEEE Workshop on Enabling Technologies Infrastructure for Collaborative Enterprises, Morgantown, WV, April 20-23.

Mahling, D.E., Craven, N. and Croft, W.B. (1995) From Office Automation to Intelligent *Workflow Systems*. IEEE Expert.

REFINE-Proposal (1995) Reengineering Insurance Business, *4th EU-Programme for IT*, Brussels.

Bocionek, S. (1995) Agent systems that negotiate and learn. *Int. J. Human-Computer Studies*, No. 42, 265-288.

Silver, B. (1995) Selecting a Workflow Tool, in: *Proceedings of the international conference WORKFLOW '95 - Busines Process Re-Engineering*. Boston, MA, March 8-10.

Workflow Management Coalition Members (1994) Glossary - A Workflow Managment Coalition Specification.

Yu, E. and Mylopoulos, J. (1994) From E-R to "A-R" - Modelling Strategic Actor Relationships for Business Process Reengineering. *13th International Conference on the Entity-Relationship Approach*, Manchester, United Kingdom, December.

Part 2 Business Process Modelling and Engineering Processes

Implementation of Information Systems Supporting Engineering Processes Based on World Wide Web

H. Grabowski, M. Furrer, D. Renner, C. Schmid

Business Process Management for Open Processes: Method and Tool to Support Product Development Processes

St. Berndes, A. Stanke, K. Wörner

Distributed Co-operative Modelling of Production Systems

B. Scholz-Reiter, D. Bastian

Implementation of Information Systems Supporting Engineering Processes based on World Wide Web

Hans Grabowski[1], Dirk Renner[2] and Claus Schmid[3]

[1]Institute for Computer Application in Planning and Design, University of Karlsruhe, Kaiserstr. 12, D-76131 Karlsruhe, Germany, gr@rpk.mach.uni-karlsruhe.de
[2,3]Computer Science Research Centre at the University of Karlsruhe, Haid-und-Neu-Straße 10-14, D-76131 Karlsruhe, Germany, {renner, schmid}@fzi.de

Abstract. The description of a Quality Management Systems referring ISO 9000 is usually documented by a QM manual, which defines for example the QM Policy, the QM System or the quality oriented business processes. The situation, the revision or the availability of the information, documented by the QM Manual, are only some of the problems, which can occur during the usage of a paper-based QM Manual. The following article describes a new approach, which combines a tool for enterprise modelling and a Product Data Management System (PDM) with the Internet-technology World Wide Web (WWW) for generating a company-wide Information System (so-called Intra Net System).

Interested readers note a realistic example which is provided at the Internet address http://www.fzi.de/divisions/cad/projects/iskoq/iskoq_demo/start.html. At this address an interactive example is given, showing the first prototype of the announced ISKOQ-Systems.

Keywords. World Wide Web, Business Process Reengineering, Data and Process Modelling, Intra Net, ISO 9000, Quality Management, PRISMA-Tool

1 Introduction

Since 1995 the of Computer Science Research Centre at the University of Karlsruhe (FZI) was been involved in the ISKOQ-Project (Information and communication system for the Quality Management), which was motivated by a study analysing the connection between a deficient supply of information within several companies in the early phases of order processing and the resulting production costs and time-to-market. The outcomes of this study demonstrated, that the correct, directed support of processes with information reduces the late growing of costs for the removal of any defects.

During the ISKOQ-Project, which is initiated and sponsored by the German Ministry of Education and Research, order-oriented companies of the production sector (e.g. Mannesmann Demag AG, Duisburg) work together with a research centre (FZI) and a software development organisation (Eigner + Partner, Karlsruhe). The project aims to the design, the implementation, and test of an information system (IS) for Quality Management (the ISKOQ-Systems). The most important goals of the IS are

- to provide the employees with up-to-date and correct quality-related information in a user-friendly and location-independent manner;
- the support of communication paths between the staff;
- the conception of an IS with client/server-architecture that enables the integration of any software-applications and any kinds of documents or information;
- the implementation of an IS with high technical performance, low costs in training and maintenance and with a maximum of acceptance by the end-user (e.g. by low training, clear functionality, up-to-date information and easy installation of the software);
- world-wide access by external employees (home-workers), suppliers (the suppliers' integration into the order processing) and customers (marketing, new services).

The project has started in October 1994 and is divided into four periods: business process and requirement analysis, conception and design of ISKOQ systems, implementation of the IS and finally the installation and verification of the IS. The following article describes each phase of the project.

2 Business Process and Requirement Analysis

At each involved company the projects have started with the analysis of their business processes that have to be performed within the context of order process·   (see also Fisher (1996), Becker (1995), Bullinger (1993) and Kagermeier (1996)). This approach reflects the experience that most of the faults made during the order processing resulted from missing information (organisational and technical) as well as insufficient information flows between the company on the one side and customers and suppliers on the other side.

In a first step all core operating processes and their supporting processes have to be identified before describing each of their value-adding activities. The second step was fixed by identifying the data input and output for these processes (the data or information flow). In the third step all personal and technical resources used to produce the process result were documented.

The analysts' most important information source for getting the knowledge about the processes, the data flows (used by activities) and the resources (for the transformation of the process-input into an output) were given by the exploitation of procedural instructions as well as many interviews with the employees who operate the processes. To ensure a structured documentation of the analysed business processes and organisation structures the staff of the project accepted the necessity to use an integrated modelling technique out of the area of enterprise modelling.

2.1 The PRISMA-Tool

The textual and graphical documentation of the business processes   and organisation structures was realised by using the PRISMA-Tool, which the FZI in Karlsruhe has been developing since 1989 (see Grabowski (1992), Grabowski (1994), Grabowski (1995) and Kagermeier (1996)). The PRISMA-Tool is used for building up an integrated information model of a company and is distributed as a non-commercial software-prototype. The most important advantages using this tool are given by the computer-aided modelling, the guarantee of model consistency, redundancy-free junction of different views and the simulation of dynamic processes under varying conditions. For the development of an IS, integrated modelling tools can be used to support all development-phases without information breakpoints at each end of a phase (Horn, 1995). This approach results in lower costs (Eversheim, 1993) and duration of projects (Krzepinski, 1993) or time-to-market.

Fig. 2.1 illustrates the architecture of the PRISMA-Tool (see also ESPRIT (1989), Scheer (1992), Ferstl (1995), Meitner (1996) and Spur (1993)), which mainly integrates different views (data-, function-, organisation- and resource-view) into a single, coherent view (the process view). The architecture is divided into a dynamic and a static part. The dynamic one defines the parallel, sequential or alternative steps to work the business process activities to the end. The relations and constraints between the objects of the same class (e.g. data, functions,

112

organisation-units, roles, persons, hard- and software) are represented by different static views.

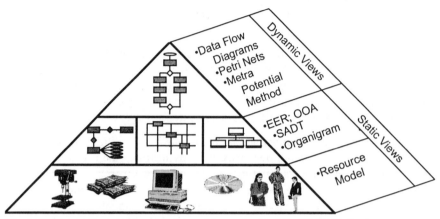

Fig. 2.1: The architecture, the modules and the methods of the PRISMA-Tool

Each view of the integrated model is defined by a well known modelling method. For the documentation of business processes the analyst can select alternatively two existing methods, either Extended Data Flow Diagrams (EDF, see (DIN, 1983)) or Petri Nets (PN, see (Abel, 1990)). The advantages of Extended Data Flow Diagrams are determined by a high transparency and easy usage, Petri Nets enable the simulation of defined processes (for yielding information about costs and time). For data modelling an Extended Entity Relation Editor is provided (EER, (Vossen, 1991)) or Object Oriented Analysis Technique (OOA) defined by Coad/Yourdan (Coad, 1991)) can be performed. The function model can be described by the Structured Analysis Design Method (SADT, see (Marca, 1988) and (Strack, 1988)) and/or simple Function Trees. Objects of the organisation (e.g. employees, departments, roles) are represented by organisation-charts (Schertler, 1993), (Schwarzer, 1995).The relations of technical resource elements (e.g. machines, storage, soft- and hardware) will be modelled in PRISMA-owned notations.

The ISKOQ-Projects started with the documentation of the organisational structure of each of the involved companies. The existing organisational units or departments, their mutual (backward) relations, the positions within each of this units and the persons, representing these positions, are the most important components or objects for the description of an organisation view. Based on the responsibilities, competencies, tasks and authorities the PRISMA-Tool differentiates between three organisation units (see Fig. 2.2).

An organisation unit which owns a personal authority and technical competence is described by a line-position (symbolised by a rectangle symbol). Line-positions are mutually connected by bi-directional relations. These relations, represented by a graphical line from a first unit (the source of the line) to another unit (the destination of the line) defines the hierarchical order of the line-position (the object at the source of the line is the super-unit of the lines' destination object).

The plain line-position organisation can be extended by a staff office (symbolised by a rounded rectangle), which neither obtain any leading task nor are they the last operating positions of an hierarchical organisation structure or a sub-unit of a line-position. The competence of a staff office lies in the preparation of decisions, the controlling and the general technical advice of line-positions. A staff office has no authority in contrast to any other organisation unit.

While these two organisation units define the static structure of a company, the time-related existence of an organisation unit is represented by a so-called project-unit (symbolised by an angular rectangle). A project-unit will be created for example, when a company works on an order with employees from different business units during a well-defined period.

Each of the organisation units has to carry out various activities, which stem from the process view. Each unit owns at least one boss and one employee. Their tasks, responsibilities, authorities and qualifications are defined by the corresponding position descriptions.

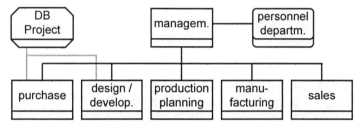

Fig. 2.2: Example of the organisation view

For analysis and documentation of business processes during the ISKOQ-Project the industrial partners were using Extended Data Flow Diagrams. It was intended to use a graphical method to guarantee a better understanding of the documented processes even by unqualified employees.

Business processes are defined by activities (rectangular symbols) or sub-processes and their sequences. To transform an input into an output (e.g. information or material), processes need technical and personal resources (e.g. machines, organisation units, employees). Each business process has a defined start as well as a defined final point (Neuscheler, 1995).

The following figure (Fig. 2.3) shows a simplified example of a business process designed with the PRISMA-Tool in the graphical notation of Extended Data Flow Diagrams. The order processing begins at the start-symbol. Each activity receives an input (e.g. inquiry, order) and is able to produce an output (e.g. the offer in process). Organisation units are in charge of the transformation process.

114

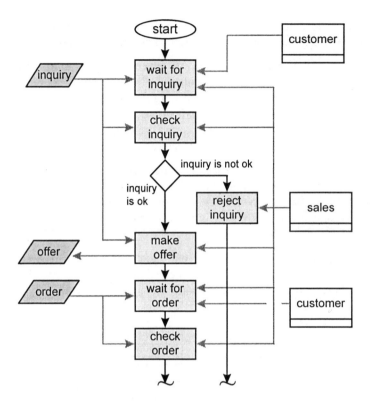

Fig. 2.3: Example of a business process modelled by the PRISMA-Tool

The technical resources used (Ciesla, 1995) are not displayed in Fig. 2.3. Like the symbols for organisations units the displayed symbols for information (e.g. inquiry, offer, order) are pointers which refer to their originals in the information view. Based on the identified information inputs and outputs of each activity, the user is able to generate stepwise the information model. The function model that is not further explained refers to the defined and described activities of designed business processes (see also (Marca, 1988) and (Strack, 1988)).

The information view focuses on the underlying data model (data elements, their attributes and relations). While the process model is used to define which activity needs which kind of information-input to generate a new information or to change information (into an output of the activity), the information view shows only the structures and the relations of the information.

In a conventional manner the data model represents the basis for the development and implementation of an IS for Quality Management (Pfeifer, 1993). Fig. 2.4 shows an reduced example of an information model. It contains the objects (entities) inquiry, offer and order, which are used as input and/or output information in the order processing of Fig. 2.3.

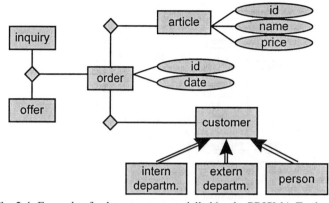

Fig. 2.4: Example of a data structure modelled by the PRISMA-Tool

In order to achieve a reduced complexity of an enterprise model (which yields a better handling for the user), the significant objects of the real world and their relations will be divided into several areas (model-views). To be able to re-use one object several times in different contexts, the objects have to be defined one-time in one of the static models (e.g. data-entities in the information view, functions in the function view, organisation units in the organisation view). Inside the process view the objects out of the different partial models will be joined together for reaching the goal of a coherent business process.

As Fig. 2.1 has shown in an abstract manner, a business process integrates the various objects from several views (or partial models) into one coherent process. Based on the previous figures Fig. 2.5 demonstrates the relations between the process objects and their originals in the static views.

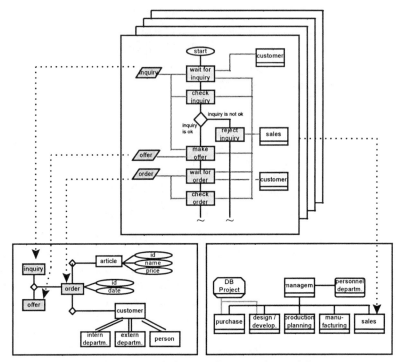

Fig. 2.5: Example for the integrated views on the enterprise model

2.2 Requirement Analysis

The textual and model-oriented description of the analysed business processes marked the end of the first project-phase in the ISKOQ-Project.

Some weeks after starting with the first phase of the project it has been found out that people's understanding of company-wide processes was already growing:

- most of the mistakes in the past will be prevented by using check-lists and weaknesses-descriptions for each value-adding activity;
- for each activity the analysts have collected all available papers, catalogues, descriptions or documents to get a survey of the necessary information needed to transform the activity in an efficient and quality-oriented way;
- those people who are involved in the documented processes received more information on responsibilities, competencies, authorities and other organisational connections. From now on, their mutual communication reduced the misunderstandings of the past;
- the analysis of process weakness, performed at the same time as the general process analysis, pointed out a lot of opportunities for continuous optimisation. This refers to the idea of the Total Quality Management (Stauss, 1996).

It was not only the analysis of the business processes referring to order processing but also the comprehensive analysis of the IS' requirements, which was one of the most important tasks in the first phase of the project. Based on the results of the process analysis the requirements for the ISKOQ-systems were derived, guarantee

- economic installation, maintenance and usage of the ISKOQ-Systems for small and medium-sized companies;
- integration of all available data (e.g. orders-, project-, customer- or supplier-data), software-applications (e.g. text-processing, spread-sheets, graphical design) and complex software-systems (e.g. Product Data Management Systems) into the ISKOQ-Systems;
- high performance, user-friendly interfaces and usage of standardised software for the implementation of the ISKOQ-Systems;
- the ability to manage different kinds of multimedia information-types for the description of static structures and dynamic processes of the company (e.g. texts, images, videos and sounds);
- company-internal and world-wide access on the information being present in the ISKOQ-Systems;
- distinction of order-dependent information (e.g. engineering drawings, parts lists, project-information) from order-independent knowledge (e.g. how to process the order)

3 Conception and Design of ISKOQ-Systems

The given user-requirements for the ISKOQ-Systems, the low acceptance of market-available systems by the involved companies and the necessity to transform the modelled business processes into any running application reduced the possible amount of solutions.

On the one hand, high expenditure was made by the companies involved for administration of information coming up during the order processing. The continuous modification of information is order-dependent and will usually be processed by different applications along the process chain starting with the order-receiving and ending with the delivery of the agreed product or service. In the early phases of the product life cycle these kinds of information are managed by Product Data Management Systems (see (Eigner, 1993) and (Grabowski, 1994). Consequently the ISKOQ-Systems have to be capable of supporting the administration of the order-dependent information as well as processing these information. Because of the multiple access on these information by different users at various places (inside and outside the company), these information will be changed frequently and/or dynamically. Based on this situation the so-called Dynamic ISKOQ-Systems (DIS) were defined for representing all order-dependent information and their applications (see Fig. 3.1).

On the other hand, the product quality and price results from the process know-how, which the company and their employees owns. This know-how exists not only by qualification of the employees, it is also defined by several documents, describing how, where and at what time someone has to do a special task (e.g. documents describing the responsibilities, authorities, activities, processes, organisation, reports, news). In contrast to the information of the DIS, this information keeps its validity for a longer time. This static nature of order-independent or organisational information resulted in the Static ISKOQ-Systems (SIS), which is the second part of the ISKOQ-Systems and contains all the static information or documents for the different business processes.

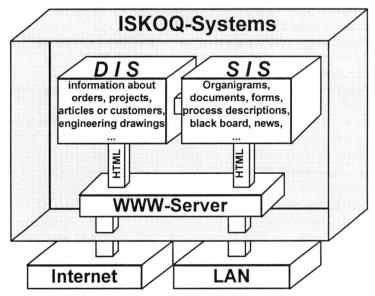

Fig. 3.1: The architecture of the ISKOQ-Systems

The distinction between the two classes of information the question was, how to present and provide these two kinds of information to the user.

The advantages of the distributed client/server-Technology World Wide Web (WWW) and its data interchange language Hypertext Markup Language (HTML) supplied a possible solution with the characteristics as follows (Morris, 1995):
- quick access to world-wide distributed data and information;
- simple use even for computer-newcomers;
- multimedia-based communication;
- world-wide structured search for information (by search-engines);
- various corners for discussion (every kind of business and private oriented topics);
- high up-to-date and redundancy-free information;
- independence from hardware;

- capability of transferring documents of every type and start them with their application;
- linking of different document types, user-friendly transferring and opening with suitable applications.

WWW-applications, so-called WWW-Browser like Netscape Navigator, Mosaic or Microsoft Internet Explorer, serve as front-end tool for access to information by means of the Internet.

After a careful comparison of the functionality of the WWW with the users' requirements on the new IS it was decided to create both the Dynamic and the Static ISKOQ-Systems basing on the HTML-format (see Fig. 3.1). The following reasons have been of prime importance for this:
- the simplicity to build up an IS based on the client-server-architecture of WWW;
- the user-friendliness to handle the WWW-applications (even an unqualified employee will learn it in a short time);
- the easy way to receive any kind of information from all over the world given by the WWW
- WWW-Server and Clients support the administration and integration of multimedia information (e.g. text, sound, picture, animation);
- integration of further information systems (e.g. Product Data Management Systems) by using the capabilities of the HTML-interface;
- WWW-applications support the integration of communication systems (e.g. e-mail);
- an incredible growing of offered WWW-services and information (in December 1994 about 10.000 WWW-Servers world-wide, in January 1996 already more than 100.000 WWW-Servers).

Depending on the users' demands HTML-documents will be distributed and accessed either world-wide by using the Internet (the WWW in real sense) or company-wide by using Local Area Networks (the Intra Net).

4 Implementation

The first phase of the project (the analysis of business processes) led to a large and integrated enterprise model documented by the PRISMA-Tool-own graphical notations and stored in binary files. The documented business processes were of great importance. They contained all information the responsible employees have to use to carry out predefined working steps. Additional for each analysed and documented working step or activity a complete description of all possible weaknesses was made to help the responsible users avoiding any faults during their work and to ensure a growing quality of their results. In order to be recognised by the users that an activity (symbolised by a rectangle, see Fig. 2.1) has linked digital documents (e.g. a detailed descriptions of the activities' proceedings, weaknesses, instructions or any forms) the PRISMA-Tool offers the functionality that the abbreviation of each linked document is displayed at the top right corner of the activities symbol. Fig. 4.1 shows the symbol of the activity <ordering material> and at the top right corner of the rectangular symbol the corresponding abbreviation (MO). This one reminds the user of the existing document that describes the way s/he has to order material. To regain the linked documents, their physical storage address is given to the activity. This address is either defined by the relative path (from the HTML-page the document is referenced) or the absolute WWW-address defined by an URL (Unique Resource Locator, starting with <http:\\www...>). The displayed screen-shots in the following figure represent the two applications needed in order to process the linked documents.

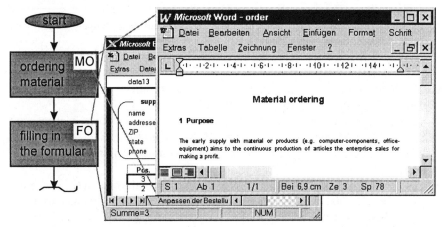

Fig. 4.1: The integration of digital documents into symbol of an activity

At the end of this extended documentation of the analysed business processes along the order processing all order-independent information were stored in the integrated enterprise model of the PRISMA-Tool, separated by different views (process-, data-, function- and organisation-view). In order to provide all the information to all their future users, is was necessary to transform the integrated model of the PRISMA-Tool into the HTML-Language.

4.1 The Static ISKOQ-Systems

Our approach was to store in a first step each graphical model-view into its own graphic-file by using the Graphical Interchange Format (GIF). According to this procedure each model view of the PRISMA-Tool consists of a graphic-file stored in GIF-format, which normally can be displayed by any WWW-browser. To avoid that all relations between the objects of the enterprise model are lost, the storing-routine of the PRISMA-Tool creates for each model view a separate HTML-File. This file includes the accompanying GIF-File and contains the information about the graphical position of the objects in the GIF-image and the links from these objects either to other objects displayed in other GIF-files or to referenced documents at any site in the Internet or at the own LAN. After the load up of this HTML-File the WWW-browser generates invisible frames surrounding each object. In case of moving a graphical cursor inside such a frame, the Browser-application recognises the referenced document and offers the user to load it (at present this feature is only implemented by Netscape Navigator 2.x). After clicking the mouse-button, the WWW-Browser loads the document and starts the corresponding application.

The mapping of the enterprise model into a net of HTML-documents enables the presentation of the business processes and their referenced documents to all employees and users of those business processes. The most important advantage of this approach is the presentation and handling of the business process models without their generating application (the PRISMA-Tool) or an expensive administrating system (e.g. a Database Management System) in a manner, the user can navigate through interactively.

4.2 The Dynamic ISKOQ-Systems

Another result of the business process analysis was the description, which kind of information object is needed as input for an activity and which is changed or generated by an activity. In contrast to the process views the information model (displayed by the information view) contained the information objects, their attributes and their mutual relations. The designing of the information model using the Entity Relationship Method of the PRISMA-Tool was needed to adapt the information model to a data management system (e.g. Workflow Management or Product Data Management System). Consequently, a WFMS or PDMS could be used to support the analysed business processes either by suitable workflows or by the administration of the needed information (Eigner, 1991). CADIM/EDB, a software-product of the project-member Eigner + Partner GmbH, was configured corresponding to the business processes.

In the past the only way to access the information managed by the CADIM/EDB-Server was to install a client application at each computer which is connected with the server-process during the whole session. The architecture of the future ISKOQ-Systems however required the integration of both parts, the SIS and the DIS into one framework, which is based on the WWW-technology. If a user wants to be informed about any organisational contents (e.g. how to process a special task), he will need to access the relevant information without handling two or more different applications. To receive all information by means of one application, access to the data of the PDM-System has to be executed by the

WWW-Browser. Therefore, prepared HTML-forms can be used to start an inquiry to the PDM-Server (see Fig. 4.2). Based on CGI-Scripts (Common Gateway Interface) the inquiry received by the WWW-Server from any WWW-Client (e.g. an customer, supplier or employee) will be forwarded to the server-process. The server processes the received inquiry and sends back the results (in form of an HTML-Document, with links to other information or documents if needed) to the WWW-Server who supplies them to the WWW-Client.

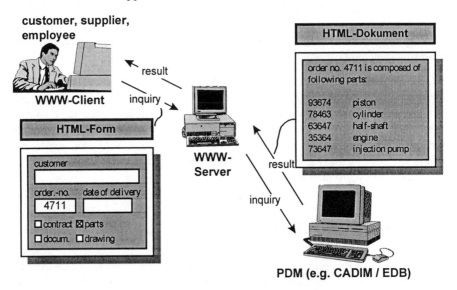

Fig. 4.2: The integration of a PDM-System into the ISKOQ-Systems

To find the right HTML-forms without an awkward and involved search, all these pages can be reached by a simple link in the graphical descriptions of the business processes. Additional almost each WWW-Browser is in charge of the bookmark-functionality that stores the addresses of HTML-pages on the users' demands.

4.3 Security of ISKOQ-Systems

Security aspects concerning the use of the SIS and DIS by authorised employees are solved using the following strategies:
- The information of the ISKOQ-Systems will be provided only for employees inside the company by using a LAN. This case needs no access to the Internet. Individual access can be realised by security features provided by the WWW-Server (in case of SIS) or the accessing-authorities managed at the PDM's side (in case of DIS).

- Opening a gateway to the Internet enables access for all employees to important and helpful services and information as well as an uncontrolled and time-spending surfing. External access to the information of the ISKOQ-Systems is generally no problem, it can be prevented however. Control is enabled providing public access to general information and restricted access to private information. For this, the WWW-Server who manages all interaction, stores the IDs of all legitimated WWW-Clients.

4.4 Example of the ISKOQ-Systems

After the generation of the HTML-Pages for the Static and Dynamic ISKOQ-Systems it has to be ensured that the concerned user can find the needed information straightforwardly. Some WWW-Browser (like Netscape Navigator) provide functionality to divide the window into several sub-windows, so called frames. The frame on the left side of Fig. 4.3 is used to display the table of contents of all available information in the ISKOQ-Systems. This ensures an all time survey about the offered information and easy navigation through the IS. If the user selects one of the items (e.g. the home-page, the first item in the left frame), the Browser displays its contents in the main-frame at the right side.

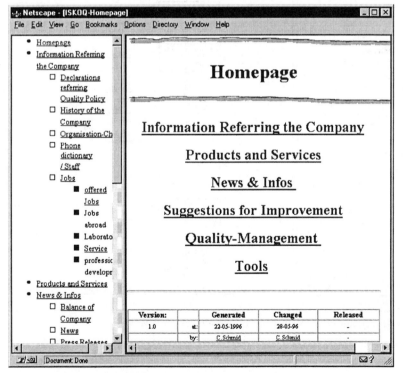

Fig. 4.3: The Home-Page of the ISKOQ-Systems

124

Each HTML-page contains at its bottom line some information about its actual status (e.g. version, date of generation, last change). Furthermore, the person's name who is responsible for the contents of this page, is shown. With a simple mouse-click the user receives a standardised form for sending any information to the named person (e.g. suggestions for improvement, new documents).

The following picture (Fig. 4.4) shows in its right frame one of the documented business processes. After analysing and documenting with the PRISMA-Tool, it has been automatically transformed into WWW-legible HTML-format. Generally, the graphical process informs the user about its single steps and the responsibilities of different departments. Supposing that the user needs any more information to a special working step, he only has to click with the mouse-pointer on the graphical symbol of the activity or the icons of its referenced documents. When the referenced document is a HTML-file it will be displayed by the WWW-Browser otherwise the WWW-Browser opens the document with the suitable application (see Fig. 4.1).

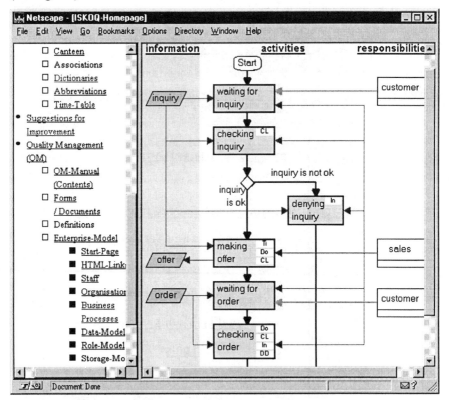

Fig. 4.4: Example of an Interactive Business Process in WWW

5 Installation and Verification of ISKOQ-Systems

The installation of the ISKOQ systems is subdivided into several steps. First of all the WWW-Server has to be installed in the LAN of the company. Simultaneously, the future users have to be determined. Normally, the analysis and documentation of business processes does not concern the whole company, but only some organisation units along the business process. Correspondingly only those employees involved in the implemented processes need access to the information in the ISKOQ-Systems and consequently the installation of a WWW-Browser. It is of great importance to maintain a running suggestion-system for improvements enabling the employees to insert their ideas into the growing ISKOQ-Systems.

The verification of the ISKOQ-Systems takes place by watching the use of the ISKOQ-Systems. Fuzzy criterions (e.g. user-acceptance, optimised communications, transparency of business processes, higher motivation of the employees by growing understanding of department-overlapping processes, easy access to information and documents, known organisational areas of responsibilities) have to be checked as well as crisp criteria (e.g. reduced time-to-market, process-costs, less quota of faults).

6 Discussion

The transformation of dynamic and static enterprise-models into HTML-documents enables their presentation in the World Wide Web, whereby all business process modelling systems can benefit from the advantages of multimedia-oriented networks and applications. This leads to world-wide or company-internal access to information as well as easy usage of consistent and up-to-date process descriptions. Furthermore, all documents for the processing of any activity can be included into and provided by graphical process descriptions via WWW. By means of the WWW the processes are distributed to all concerned employees along the examined process chains without special and expensive systems.

For future steps it is planned in the depicted Project (ISKOQ) to create new facilities for the direct access to applications supporting the business processes (e.g. Product Data Management Systems) and to their administrated information (e.g. bills of material, order-data, engineering designs) by means of a WWW-Browser.

The key for yielding more benefit with Business Process Reengineering will be received by the future multimedia capabilities of hard- and software as well as the growing performance of world-wide networks. In a transparent company-policy manner the integration of multimedia documents becomes of prime importance (e.g. for public relation, training of employees, pictured manufacturing processes).

126

The Information Systems described in this article will be realised within the ISKOQ-Project. This project is financed by the German Ministry of Education and Research and is part of a project, which is focused on Quality Management. The results of the ISKOQ-Project aim to small and medium sized companies.

References

Abel, D. (1990): Petri-Netze für Ingenieure: Modellbildung und Analyse diskret gesteuerter Systeme; Springer Verlag

Becker, J.; Rosemann, M.; Schütte, R. (1995): Grundsätze ordungsmäßiger Modellierung; Wirtschaftsinformatik; Friedrich Vieweg & Sohn Verlagsgesellschaft; Braunschweig / Wiesbaden, 5/1995

Bullinger, H.-J.; Fröschle, H.-P.; Brettreich-Techmann, W. (1993): Informations- und Kommunikationsinfrastrukturen für innovative Unternehmen; Führung und Organisation; Fachverlag für Büroorganisation und Organisationstechnik; 4/93

Ciesla, M.; Gräßler, R.; Ordenewitz, R. (1995): Integriertes Ressourcenmanagement; ZWF, Zeitschrift für wirtschaftlichen Fabrikbetrieb; Carl Hanser Verlag, München; 3/1995

Coad, P.; Yourdan, E. (1991): Object-Oriented Design; Prentice Hall, Englewood Cliffs, New Jersey

DIN, Deutsches Institut für Normung e.V. (1983): DIN 66001 - Informationsverarbeitung: Sinnbilder und ihre Anwendung; Beuth Verlag, Berlin, Köln

Eigner, M. (1991): Engineering Database - Strategische Komponente in CIM-Konzepten: Hanser-Verlag, München, Wien

Eigner, M.; Rüdinger, W. (1993): Betriebliche Einführung von EDM-Systemen; CIM-Management; R.Oldenbourg Verlag; 5/1993

ESPRIT Consortium AMICE (1989): Open System Architecture for CIM; Springer Verlag

Eversheim, W.; Krumm, S.; Heuser, T. (1993): Ressourcenverzehr minimieren durch Ablauf- und Kostentransparenz; CIM-Management; R.Oldenbourg Verlag; 3/1993

Ferstl, O. K.; Sinz, E. J. (1995): Der Ansatz des Semantischen Objektmodells (SOM) zur Modellierung von Geschäftsprozessen; Wirtschaftsinformatik; Friedrich Vieweg & Sohn Verlagsgesellschaft Braunschweig/Wiesbaden; 3/1995

Fisher, B. (1996): Reengineering Your Business Process; The Journal of Systems Management; Cleveland; USA; Jan/Feb 1996

Grabowski, H.; Schäfer, H.; Krzepinski, A. (1992): PRISMA - Methodisch unterstützte Planung und Integration von CAD/CAM-Verfahrensketten; CIM-Management; R.Oldenbourg Verlag; 6/1991, 1+2/1992

Grabowski, H.; Anderl, J.; Polly, A. (1993): Integriertes Produktmodell; Beuth-Verlag, Berlin, Wien, Zürich

Grabowski, H.; Furrer, M.; Schmid, C. (1994): Planung und Simulation von Prozeßketten unter Berücksichtigung einer schlanken Produktion; in Vereinfachen und Verkleinern - Die neuen Strategien in der Produktion; (ed. G.Zülch); Schäffer Poeschel Verlag, Stuttgart

Grabowski, H.; Krug, W.; Schmid, C. (1995): Prozeßmodellierung und Simulation im Produktlebenszyklus; Simulation in der Praxis - Neue Produkte effizienter entwickeln; VDI Berichte Nr. 1215

Horn, H.; Brockhaus, R. (1995): Workflow-Mangement- und EDM-Systeme; ZWF, Zeitschrift für wirtschaftlichen Fabrikbetrieb; Carl Hanser Verlag, München 1995; 7-8

Kagermeier, K.; Renner, D.; Schmid, C. (1996): Workflow-Management in der Kreditsachbearbeitung; Bank und Markt + Technik; 25. Jahrgang; Fritz Knapp Verlag, Juni 1996;

Krzepinski, A. (1993): Ein Beitrag zur methodischen Modellierung betrieblicher Informationsverarbeitungsprozesse; Verlag Shaker; Aachen

Marca, David A. (1988): Structured Analysis and Design Technique; McGraw Hill; New York

Meitner, H. (1996): Infrastruktur für die Geschäftsprozeßoptimierung - die Unterstützung durch Infobahnen; Zeitschrift für Unternehmensentwicklung und Industrial Engineering; REFA-Verband, Darmstadt, 1/1996

Morris, M. E. S. (1995): HTML - WWW effiktiv nutzen; Verlag Heinz Heise

Neuscheler, F. (1995): Ein integrierter Ansatz zur Analyse und Bewertung von Geschäftsprozessen; Forschungszentrum Karlsruhe - Technik und Umwelt; Wissenschaftliche Berichte Nr. FZKA 5637; Karlsruhe

Pfeifer, T. (1993): Qualitätsmanagement - Strategien, Methoden, Techniken; Carl Hanser Verlag; München

Scheer, A.-W. (1992): Architektur integrierter Informationssysteme - Grundlagen der Unternehmensmodellierung; Springer Verlag

Schertler, W. (1993): Unternehmensorganisation, Lehrbuch der Organisation und strategischen Unternehmensführung; R.Oldenbourg Verlag München Wien

Schwarzer, B.; Krcmar, H. (1995): Zur Prozeßorientierung des Informationsmanagements; Wirtschaftsinformatik; Friedrich Vieweg & Sohn Verlagsgesellschaft; Braunschweig, Wiesbaden; 1/1995

Spur, G.; Mertins, K.; Jochem, R. (1993): Integrierte Unternehmensmodellierung; Beuth Verlag; Berlin, Wien, Zürich 1993

Stauss, B.; Friege, C. (1996): Zehn Lektionen in TQM; Harvard Business Manager; 2/1996

Strack, V. (1988): Zur Semantik graphbasierter Modellierungsmethoden; Gesellschaft für Mathematik und Datenverarbeitung; R. Oldenbourg Verlag; München, Wien

Vossen, G. (1991): Entwicklungstendenzen bei Datenbank-Systeme; R.Oldenbourg-Verlag; München, Wien

Business Process Management for Open Processes
Method and Tool to Support Product Development Processes

Stefan Berndes[1], Alexander Stanke[2] and Kai Wörner[3]

[1]Chair Philosophy of Technology, Brandenburg Technical University Cottbus, Karl-Marx-Str. 17, 03046 Cottbus, berndes@tucs1.rz.tu-cottbus.de
[2,3]Fraunhofer-Institut for Industrial Engineering, Nobelstr. 12c, 70569 Stuttgart, {alexander.stanke, kai.woerner}@iao.fhg.de

Abstract. Product development processes are open and creative business processes. They consist of a project level where the coordination of different engineering processes takes place. These sub-processes form a second level. The engineering processes can be supported with business process supporting tools like workflow and document management systems. On the project-level the sub-processes must be bound together, coordinated and their execution has to be controlled. Such business process management for open and creative processes can be supported by the Engineering Process Manager (EPM) which is a generic IT-tool. The module supports a coordination method which reflects a critique of project management methods which support a central information management and decision making rather than decentral management with distributed knowledge and competence. Process management is considered as an ongoing detailation, negotiation, decision making and information process about the product development process between teams and a project manager. EPM helps to collect, disseminate and interpret data to plan, schedule and control the process along important issues like: planning, replanning and controlling. The performed process can be visualised and criticised at arbitrary times. After the product development process has been finished EPM supports the evaluation of stored data to improve and update the standards for the business process management for open, creative processes.

Keywords. Business Process Management, Concurrent/Simultaneous Engineering, Engineering Process Manager (EPM), Multi Agent Approach, Open and Creative Processes, Product Development Process

1 Introduction

The product development process is of crucial importance for most manufacturing companies. In the case of the development of complex products it is characterised by the complex coordination of a number of strongly interrelated engineering processes like mechanical design, FEM-calculations, prototype building etc. Changing requirements, technological problems and creativity disturb plans. Therefore each project realises a unique process. Is it possible to apply methods of business process management to the sphere of product development processes as typical open and creative processes, too?

The aim of this paper is to present an answer to this question in the following way: Starting from definitions of business process and business process management the management of product development processes is identified as a special case of business process management. In a next step the discourse about business process management and the discourse about Concurrent/Simultaneous Engineering, which is completely concerned with product development processes are related to each other. If Concurrent/Simultaneous Engineering could be understood as business process management for open and creative processes, research results from this area should be incorporated into the business process management discussion. In the arena of Concurrent/Simultaneous Engineering tools and methods have been developed to optimise the development process. An important part of this discussion is about project management (Bullinger, et.al., 1995). The proposed tool and method support the project management for product development processes and have been developed along a critique of current project management methods. The critique and the main concepts of method and tool are presented. In order to identify method and tool as business process management instruments they are evaluated under this aspect. The remainder of this paper is dedicated to the presentation of further approaches for business process management for open and creative processes and open questions.

Method and tool which are presented have been investigated in the frame of the ESPRIT-III Project CONSENS (Concurrent/Simultaneous Engineering System) EP 6896 by researchers of the Fraunhofer Institute for Industrial Engineering in cooperation with other European research institutes, industrial partners and users.

2 Business Process Management and Product Development Process

Business processes are an accepted concept to discuss organisational issues in enterprises (Ferstl/Sinz, 1993). The aim of business process management is to strengthen organisation's ability to achieve their goals in an environment which is characterised by growing complexity (Sinz, 1995). Following a commonly used definition a business process is a set of logically connected decisions and activities designed to produce a specified output for a particular customer or market (Ferstl/Mannmeusel, 1995 and Davenport, 1993, p. 5). Business processes do typically need input from other business processes. They follow given formal goals. The coordination of business processes can be analysed in initiation, arrangement and execution transactions (Ferstl/Mannmeusel, 1995). „Business process management is an approach of model-oriented design, coordination and execution of business processes. Its objective is to combine tool-based enterprise models [..] with enterprise wide information systems and application services. [..] feedback and change management methods must exist to involve the agents and company parties (business process owners, service centres, organisational units etc.) which are part of the managed business processes" (Scheer et.al., 1995, p. 426).

The product development process is a special business process and covers the process from the idea of a product to its market entry. It is a defined set of activities which are required to generate the necessary information about the product, its production, logistics, marketing efforts etc. (Berndes/Stanke, 1996, p. 16). Patterson (1993) uses a sound metaphor: product development is an information assembly line where information about a new product or service is gathered, created, integrated and documented. The product development process shows typical characteristics. Specht and Beckmann (1996) made the valuable effort to classify tasks along formal and functional variables. Patterson (1993) gives an analogue classification of processes. Along their variables the product development can be characterised as a high complex process. Its sub-processes are interdependent, the degree of innovation is high and the task itself seldom clearly defined. This limits the possibility to plan the whole process. The information supply in many cases is limited, necessary information is generated often later than it is needed first. Many tasks need creative, inductive problem solving methods. The aim of the whole process can not be defined exactly. In consequence the demand for mutual communication grows. Informal communications are important. Openness, unforeseeable obstacles and problems require flexibility. Product development is knowledge production, it is open. This knowledge is basis for the next product development process. Therefore no development process will be like an other. Last but not least, the number of development processes is limited. That may be another reason that up till now in most companies product development is more exceptional than routine. Of course, this characteristics vary from task to task and project to project. Nevertheless, it is plausible to suggest that there is a pattern, which leads us to conclude, that product development processes are significantly different from most other business processes - they are open and creative business processes. But a plausible hypothesis does not replace further empirical studies.

The differences between the „typical" product development process and other business processes led to a discussion in the scientific community whether business process management methods should or should not be applied to them. Picot and Franck (1995) write that some authors focus their advises on structured processes only. Then processes like product development processes are consequently out of focus. Maybe a reason for this concentration is that possible profits from re-engineering efforts used for routine processes are greater than for project type processes (Nippa, 1995, p. 54). As a second reason may serve an argument given by Ferstl and Mannmeusel (1995). They present a frame concept for the implementation of redesigned business processes. It consists of separated phases to come from a model to an actual implementation of the process. This is not quite feasible for product development processes. Here modelling and implementation of the process takes place at the same time and the process will be changed often during its runtime. Managing a product development process means to design the process, to coordinate and to execute it at the same time.

Applying business process management methods to routine processes first may be efficient. But is it effective? Davenport (1993) considers the task to conceptualise product development as a process as a major task in companies because „nothing is more critical to a firm's competitive success than its ability to develop new products and services" (ibid., p. 221). Applying the process view to product development processes (like research, design and product engineering and manufacturing engineering and logistics) is important. Since product development has gained importance for most manufacturing companies because of an increased time competition (Stalk/Hout, 1990), shift from sellers to customer dominated markets (Clark/Fujimoto, 1992) and more and complex technologies integrated in products re-organisation of the product development process was on the agenda for most companies which engage in development. This discussion can be framed with the buzz words: Concurrent or Simultaneous Engineering and project management (for an introduction into the topic see: Bullinger et.al., 1995).

What is Concurrent or Simultaneous Engineering? The differences between the two concepts are in our opinion of a more of a historical interest (Berndes/Stanke, 1996, p. 15). Concurrent/Simultaneous Engineering is a strategy and methodology to optimise the product development process. The aim is to improve product quality, shorten the time to market and reduce the development costs. Three guiding principles can be fixed: parallelisation, standardisation and integration. Parallelisation focuses primarily on shortening process times, standardisation has the aim to produce transparency and stability of processes and the integration efforts strive for quality. Necessary changes concern the organisation, processes, human resources and the product itself. Here only the most important tools or methods to implement Concurrent/Simultaneous Engineering are mentioned: project management, neutral process plans, product data management, training of personnel in team skills and product modularization. Obviously, Concurrent/Simultaneous Engineering can be understood as business process management for product development processes. Both follow the same basic idea. The process orientation is inherent - one focuses on the product development process and supports this with project management (Bullinger, et.al., 1995). Some scholars already knit both discussions together. Hammer and Champy (1995) present in their seminal book on business re-engineering a case study about the Kodak product development process which was re-engineered with help of Simultaneous Engineering. If Concurrent/Simultaneous Engineering is business process management for the product development process, and if we are interested in the support of open processes, and product development processes are a type of open processes, then one should look at the results of the Concurrent/Simultaneous Engineering discussion.

3 EPM a Tool to Apply Business Process Management to Product Development Processes

The sub-processes a product development process consists of are mostly known and standardisable. Such sub-processes are for example: mechanical design, structural and dynamically analysis (finite element-modelling) in the case of mechanical engineering and the printed circuit layout process in electronically engineering. Of course the degree of possible standardisation depends on the type of development process. These sub-processes itself are often relatively simple but they have to be iterated, they depend on each other, one processes uses the result of the other process as input information and the other way round. Last but not least, unforeseeable obstacles occur, new phenomena, never expected test results. Also typical for a Concurrent/Simultaneous Engineering conform product development is to involve employees from many other organisational units which normally work on other processes like manufacturing, purchasing, controlling, marketing etc. Summing up, standardisable sub-processes which are responsibly performed by groups or teams which may not necessarily be part of the development department and the multiple, complex interdependencies between these sub-processes suggest to divide the processes in the following way. This division has been adopted in the project CONSENS which had the aim to develop an IT-system supporting complex product development.

There is (1) a level to support the standardisable sub-processes by a workflow management system and a document management system. Workflow management systems are distributed, integrated information systems on a client-server basis which support a rule-based configuration and integration of existent software components. They allow persons involved in the process to work at different times and locations on the process (Galler/Scheer, 1995). The CONSENS platform with SiFRAME has a workflow management component for product development processes (Kessler, 1996).

Due to the fact that product development tasks are standardisable for sub-processes and open on the project level, workflow management for the sub-processes should be combined with schedule coordination and document management systems (comp. Deiters/Striemer, 1994). Therefore, a document management system is incorporated in CONSENS (Miotti, 1996). Sub-processes which are standardisable and part of almost all product development processes should be performed by specialised teams. Hess et.al. (1995) require self-organising and (Case)-teams which have no detailed process regulations on the project level. This teams, which do work typically on more than one product development process or which work on another business process are coordinated by one project manager for every project. Each team is responsible for its own set of sub-processes. Ferstl and Mannmeusel (1995) explain very realistically that the product development system is a distributed system made of autonomous teams which have common aims. But no single component has global control and no single component has actual complete knowledge about the systems state. These teams prepare decisions concerning their area, meet together with the project manager to come to final decisions. The teams do the human resource allocation for their sub-processes also. These teams have also the responsibility to improve their knowledge about their area of work and their processes. The scheduling component is EPM which should (2) consist of a method to plan and to coordinate the sub-processes. The coordination task is divided between project manager and team speakers. The project manager is responsible for the negotiations and decisions inside his project and he stands for the project. Project managers and the top management of the company have the task to solve conflicts between different development processes and other business processes.

Business process management is planning new processes and controlling them, in the case of product development this task can be divided into two parts: the management of the sub-processes and the management of each product development project. The question to be investigated in the following sections is: what to do on the project level; how to initiate, arrange and execute the necessary sub-processes (Ferstl/Mannmeusel, 1995). Summing up: Given teams with distributed knowledge and a central coordinator, the project manager who work on an open and creative product development process, how should one coordinate the teams and support this coordination? Consequently, the first thing to do is to look at project management methods and IT-tools which support them. They have been discussed in the CSE-debate for long. This critique may serve as a requirements record for a new method and supporting tool to plan and coordinate the product development process on the project level in cooperation of project manager and teams. This is one possible approach to design the project management for open and creative processes which are performed by specialised teams.

3.1 Critique of State of the Art in Project Management Methods and Tools

The critique of the current project management methods and tools which is presented here uses arguments from practitioners and scholars. It is based upon experiences and analysis of enterprises ranging from 500 to 10000 employees and in project size from designing ABS controllers to aircraft. The enterprises analysed were TEMIC and DASA (Bergner, 1996; Vilsmeier, 1996). Moreover, user requirements to EPM itself have been written by DASA and TEMIC itself.

Today IT-support for the project management in product development is mostly given by project management systems (PMS) including netplanning techniques (NPT) (Schwarze, 1990 and Möhrle, 1991, p. 122). NPT is the method of project planning. PMS have been developed for projects which consist of many sub-processes which are strongly interrelated and where the interrelations are known. As we know from our experiences in industry these project management method and tools lack a few attributes supporting the management of complex product development processes (another line of critique is given in McGrath, 1992, p. 113).

During project planning the workpackage structures are defined centrally by the project manager not by the teams which have the knowledge about their processes. All current PMS require a central resource management but companies with teams which belong to different organisational units need a decentral resource management. To make matters worse, we know from consultancy projects that the resources, mostly specialised engineers for different tasks cannot be exchanged. This is a hard condition because many companies work on a few development processes in parallel: one new product, a few change developments and there are urgent change requests and further business for each engineer, too.

PMS based on NPT do not handle inconsistencies. We assume that net planning technique is sufficiently known to the reader. NPT needs complete and consistent information about all workpackages with times, efforts, resources etc. Open relations, cycles and schedule inconsistencies lead to errors. But complete and consistent information is often not existent or hard to get. These processes for each product development are partly known only (McGrath, et.al., 1992, p. 99) and they need creativity. Further occur challenging problems in a development phase whose solutions may change the course of the whole project. And we assumed that the project manager has to coordinate teams which may be engaged in more than one process. Typically a rough plan can be constructed and the detailed coordination of work is done for the near future only. The project manager who wants to use net planning technique has two alternatives. (1) he concentrates on a rough plan and he starts to set up a set of further plans for detailed coordination. The effect is a huge amount of work to update and coordinate this not interlinked and distributed information pools. (2) he has to elaborate a detailed plan. Here the project manager often ends up frustrated. He would have to collect a huge amount of information to update the detailed global plan during the project runtime. So he works with old plans and starts to edit detailed plans, too. In both cases the consistency of information and a universal database have been given up in favour of an eclecticistic management method. Therefore a method and supporting tool is necessary which is capable to accept that information about the coordination of a project is neither complete nor consistent.

Our assumption of supporting decentral teams suggests to have distributed knowledge and distributed access to input controlling data but most PMS require central data input for efforts, cost and time. The reason is a missing multi-user ability of the PMS-software (Dworatschek/Hayek, 1990). Along with this goes the problem that PMS support multi project management insufficiently because information presentation crosswise to the projects is not supported. For project controlling it is necessary to have an handy concept for the measurement of goal attainment (comp. Möhrle, 1991, p. 122 and Bürgel/Kunkowski, 1989). Project managers of industrial product development projects state that the actual share of the planned effort which has been spent so far and the opinion of the team who is in charge to finish a sub-process are not sufficient. Further, most PMS lack a reporting module which helps to generate and deliver reports or other important project information. Last but not least PMS do not support the evaluation of project experiences which should be used to improve the project management of product development processes. Anyway, project auditing and evaluation are a weak point in European project management in general.

Summing up, it seems that NPT and PMS cannot cope sufficiently with the Concurrent/Simultaneous Engineering requirements set for the management of product development processes. Therefore a new method and tool which support the project management for open and creative processes have been developed during the CONSENS project.

3.2 Method and Tool EPM

Of course, workpackage structures, netplans, Gantt-Charts etc. are powerful means to structure, plan and control a product development process and the presentation forms are rather expressive. Project managers use such charts only as we know from a great variety of projects. If project managers use the PMS as a presentation form editor without using the variety of concepts inherent to NPT, we should stick to the presentation forms and rethink the constraints NPT has. And as suggested above we have to support the whole project management which is in an environment with decentral teams and distributed knowledge a task of the project manager and the teams.

3.2.1 Concepts of EPM

We stick in the terminology to the known project management concepts. Work-packages help to structure the whole project and form a hierarchy. Each work-package (WP) has a start and an end date. For planning process reasons this dates have two different values, a bottom up and a top down value. Each workpackage has an effort. For the effort values again a top down and bottom up value is stored. The bottom up value equals the sum of the efforts of the sub-workpackages below the workpackage in question.

Each workpackage has one intermediate result (I_Res.) which defines the goal of the task. For each of them a responsible person must be named. Each intermediate result may have a couple of status such as first draft and second draft before the final status „ready" is achieved. For every status a date of delivery is defined. The date of delivery for the ready status is typically the bottom up end date of the workpackage. Intermediate results serve as mean to control the project progress. As far as we know, the best thing to control the state of work without the problems mentioned above is to build up customer-supplier chains in which the customer accepts or rejects an intermediate result on a planned date in a defined status. In order to hand over the intermediate result to the group of persons who rely on it from the point of view of the deliverer a list of customers of this intermediate result for every status is stored. Also for producing each intermediate result a list of incoming intermediate results is necessary. They are written in the list of suppliers. The planned dates here are expressed relative to the end date of delivery of the ready status of the intermediate result. The actual end date of a workpackage equals the actual date of delivery where the affiliated intermediate result is delivered and accepted. In conducting this procedure a dually interlinked intermediate result network is defined which is a schedule, a clear responsibility chart and a presentation which expresses information about the big customer-supplier network which is the product development process. This intermediate result network represents an event controlled process chain (Scheer et.al. 1995). Delivery of I_Res. and acceptance actions are events. In the words of Ferstl and Mannmeusel (1995) the approach to focus on intermediate result exchanges between the decentral teams on the project level can be understood as to hide the function and to put the result and aims into the foreground. The business process is represented as a sequence of intermediate result exchanges. How this results are worked out is not subject of project management. The functions are subject of the sub-process level which is partly supported by the workflow management component of CONSENS.

In order to control a project regarding global questions as do we already finish in time, are their global problems, has the market changed entirely or has the resulting product a good market position, the concept of milestones (MS) is very well known. Each milestone defines the end of a major development phase (specification, design, prototype testing, series start). A milestone is defined by a date, budgeted efforts or costs which can be spent during the period in question. This costs or efforts equals the sum of the efforts of the workpackages which belong to the phase. All those workpackages belong to the phase in question which are between the preceding milestone and the milestone of this phase. The milestone is defined also by results which should allow the evaluation of the project. Each result can be related to a list of intermediate results. The milestone is reached if all results are there. But in many practical cases not all results are there in time. In order to go on with the project it is possible to postpone i.e. to relate some results to later milestones.

3.2.2 Planning and Controlling a Product Development Process

The project planning and control methods are described along the functions: planning, replanning, control and audit. The idea of the EPM planning method is as follows. We are searching for the best or optimal project plan but we know some parts only. It isn't and feasible to carry together all information which are stored decentrally. Therefore we distribute the planning tasks onto the teams. The teams are not only resources, moreover, they have know how, they act as problem solvers, they consist of acting persons to overcome organisational and system-technical obstacles and barriers (Scheer et.al. 1995). We arrange discussions between the teams, classify the topics of discussion and accept inconsistencies which result from the complexity of the task, not knowing of vacation, illness, resource shortages etc. This is a multi agent approach (a similar approach is given by McGrath, 1992, p. 113). Similar for Ferstl and Mannmeusel (1995): business processes are autonomous components of a distributed enterprise system. Instead of a global control they call for decentral coordination of result relationships between the processes. As a component of the distributed system regulators like project managers and the executive level also do not have an actual, complete and consistent picture of the system as a whole. The problem in managing product development processes is to coordinate those inconsistent and incomplete world views, and to let them develop.

The proposed method of planning follows a recommendation of Malik (1986). He describes two approaches of project management. A technomorphic approach and a system-evolutionary. The second one uses organisms as paradigm of understanding. The whole development process has a certain self-dynamics. The task of the management is to lead this dynamics into a required direction. The technomorphic approach considers the project as a machine, a determined production process. The task of project management consequently is to detail tasks, to distribute and to control their results. This seems to be the philosophy which underlies NPT. We follow the organism model because the here presented method

138

should be applied to open and creative processes, where the technomorphic model has disadvantages due to the lack of knowledge.

The project manager defines the project with the major goals: quality, time and cost and inputs them in EPM. He defines the project organisation with the involved teams. He builds up a hierarchical workpackage-structure with maximum efforts for every WP. This are the top down efforts. EPM supports this by help of a workpackage-editor form. A list of milestones with explanation is selected from standards and just distributed on the time schedule. The project manager delegates the planning tasks to all involved teams. EPM sends a message to all teams. The teams refine and elaborate interactively with EPM the project goals. They know about resources and technical or other risks. They make a statement to the main schedule and the cost frame. They consider who will be assigned to the project. Each team builds up its WP-structure, using decentrally stored WP-structures. In order to use experiences from former product developments each team stores itself information, concerning WPs, typical time frames, efforts and some explanations concerning resources. They define their planned efforts, the bottom up efforts. The teams perform a rough planning of the schedule for their WPs. Without any relation to other WPs the teams distribute their WPs on their capacity chart. The sequence of WPs is fixed by their position on the capacity chart. There is a pick´n drop mechanism implemented in EPM to distribute the WPs and by pulling and pushing the WP the start/end date can be selected.

The information about the projects and the efforts each team has to deliver is stored in the capacity chart of EPM. In the capacity chart are the available resources presented along the time-axis. To come up with the capacity chart each team defines a 100%-line and a line for day-to-day work. On the basis of the planning and actual values for efforts an effort line for each project is computed. All efforts for WPs of one project phase for each project are added and this leads to a phase effort of each team. This effort will equally be distributed among the phase time. The planned efforts are defined as the efforts on which has been decided on - in terms of the WP-definition it is the bottom up-value. How can one get actual dates concerning efforts? Every week engineers have to input in EPM which project they are working on. This information is put into the capacity chart. Efforts for WPs which have not been ended during the phase runtime are put in the next phase. Due to the fact that project employees do not input effort values on the level of workpackages the equal distribution of efforts over a phase is justified.

Project managers and the team speakers evaluate the resulting plan. They regard project run time, project efforts and resource bottlenecks. Meetings are held between the team speakers to discuss and to solve appearing problems and meetings are held with the top management to prioritise different projects. Here EPM supports the generation of info charts like a milestone schedule, a WP-plan with efforts, dates and effort/costs distribution for the teams and the project. Minutes and decisions can be stored in EPM to document the decision process. This procedure can be repeated until a suitable rough plan is found. The aim of this planning phase is: generation of a stable project plan, definition of milestones (MS) and schedule, check of the general ability to realise the project.

Then the teams define their intermediate results and relate them to the MS-results. Each workpackage has a bottom up end date and this date equals in the first iteration the I_Res. final delivery date in the status „ready". If the intermediate result bottom up end date is later than the MS-date there is a conflict. This is computed by EPM automatically and solved in talks between the project manager and the team speakers. The teams build up the intermediate result network next. Each team asks which intermediate results are necessary to elaborate their I_Res. and relates them to it and ask who relies on their intermediate result. EPM supports this by storing the relations to intermediate results from former similar projects and printing lists of intermediate results of all other teams. Of course, in the interlinked intermediate result network a 'few' inconsistencies may occur. EPM helps to discover them. The following classes of inconsistencies are found. Open relations make the teams and the project manager aware of coordination problems. Information exchanges are not clear. The solution of such problems is rather simple: What does the relation mean? Why do (not) you expect an information flow? Which relations should be added or resolved? Storing the relations in the I_Res. lists make the results of this discussions available for future product developments. Cycles are often found in engineering. One team relies on the result which another team elaborates on basis of the result of the first team. A method to resolve them is to establish a sequence of smaller steps, by exchanging draft-information. Schedule inconsistencies are easily found if the I_Res. planned delivery date of the to be delivered intermediate result is later than the planned date of supply of the other intermediate result. The first thing one can do is checking whether the necessary I_Res. must be delivered on that date. Next, one may try to break down the to be delivered I_Res. into smaller ones and in doing so define a few further sub-workpackages. Or just try to exchange draft-information. A last solution is to postpone the intermediate result and the workpackage.

After discovering the inconsistencies with EPM the task of project team meetings is solving them. Project manager and team speakers work out alternative solutions on EPM by inputting and storing them separately and evaluation of this alternative scenarios, discussing them and putting in the result in EPM. The idea is to solve the appearing inconsistencies one after another. Main inconsistencies are solved first and the others are solved at the time when it is possible to foresee the near future. The result network must be consistent for the near future only - consistency just expresses good coordination between the teams. Of course, for each kind of product development and company the time frame of consistent coordination is different. Trying to have a whole consistent intermediate result network is very expensive and needs much effort to adapt to daily changing conditions but it is required from NPT.

In the terms of Ferstl and Mannmeusel (1995) the coordination of aims and result exchanges between objects of the enterprise system are hierarchical control and non-hierarchical negotiation. The executive level and the project manager set the aims using their hierarchical position and, more important, moderate initiation, arrangement and execution transactions between suppliers and customers of results or clients and servers. The main difference between netplanning technique and our proposed planning method is the introduction of intermediate results. Netplanning technique assumes that resource planning and scheduling of the project progress can be done by help of workpackages only. We differ here. In creative and inductive processes the connection between results and effort is not known exactly. Therefore we introduced intermediate results. The schedule is controlled by the intermediate result network and the effort spent by the workpackage hierarchy and this hierarchy is used to get a basis for a sufficient estimation of efforts.

The replanning procedure will be done at each milestone or any other moment wherever it seems to be necessary. For example if the planned and actual effort numbers differ significantly from each other. In order to be able to control the process both the new and the older planned numbers are stored. Replanning means just to use the same method as in planning. Smaller inconsistencies in the schedule and the intermediate result network may be handled by the teams on their own, only a short message is given to the project manager. But the decisions are documented in EPM.

Project control contains data acquisition, distribution and evaluation of data. The time-dimension is measured by comparing planned bottom up and actual data. The cost-dimension is measured by comparing planned bottom up and actual data for efforts and other project costs. And the quality-dimension is measured by help of the established customer-supplier chains through the I_Res. network. The quality of I_Res. is checked by the customers of the specific I_Res. who accept them. This is process control on a detailed level. The global project progress is controlled at the milestones by global reviews. Between each milestone the intermediate result exchange, effort spent and schedule progress can be measured at arbitrary points of time.

EPM as an information system presents data for different user classes (similar McGrath, 1992, p. 113). The project manager needs information concerning schedule, effort, cash flow and quality, because he is responsible for the project. The team speakers must know about effort tables and schedule situation. The project employees have to know where they are involved and what they have to do. At least the top management wants to have information for setting project priorities. In doing project control, problems can be found and will automatically be presented. EPM supports also agenda setting and the management of project change requests by presenting, mailing and documenting of specific forms to the involved persons.

3.2.3 Project Auditing with EPM

In doing so, for each project a tremendous amount of data is processed and stored which serves as an evaluation base. The evaluation is done in audits which are used to reflect the problems. Changes in MS, WP-structures and the I_Res. network can be proposed. EPM delivers the necessary information for the audit. Especially it is possible to print out the actual followed process. And EPM offers a way to implement changes by changing the standards MS-lists, WP-structures, I-Res. lists etc. This is support of process change management for open and creative business processes required by Scheer et.al. (1995). Nevertheless there remains the most important task for the management to really give priority to audit and process improvement meetings. The management and the team members must understand that all the data which are processed during the runtime of product development processes are not knowledge about the product development processes. Of course, these data are relevant, but knowledge depends on people who are able to interpret the data and who are able to decide on the basis of this data (Lyotard, 1993; Poser, 1990). Here we ask: who does the knowledge management in companies.

The most important task for multi project management is resource distribution between different 'competing' product development processes and establishing a preference list of projects and other processes. EPM supports a decentral resource management approach and supports in doing so multi project management. Each team has the task to administer its own resources. The results of the coordination are documented in the schedules. The task of deciding on preference lists is done by the top management. EPM here supports multi project management here by giving the top management access to comparable, true and full information on the project and its progress.

4 Conclusion

The discussion of Concurrent/Simultaneous Engineering has been fixed as business process management for open and creative processes. As long as open and creative processes are not widely discussed in business process management the results from the Concurrent/Simultaneous Engineering discussions should be introduced into the business process management discourse. A first attempt has been made here. A division between process levels, a sub-process level and a project level on which the sub-processes are coordinated and controlled showed to be fruitful. On the level of sub-processes workflow management and document management systems can be used. We focused on the project level. This is the level on which the openness of product development processes has the greatest impact. Here a new method could be helpful which was developed in the realm of the Concurrent/Simultaneous Engineering discourse.

EPM and its underlying method have been proposed as suitable means to support the management of product development processes. The problem tackled was the following: Given teams with distributed knowledge and a central coordinator, the project manager, which work on an open and creative product development process, how should one coordinate the teams and support this coordination? We fixed the problems of project management methods and supporting IT-tools which have been discussed in the CSE-debate. This critique served as requirements record for a new method and supporting tool. The main critique points are repeated and asked whether EPM is able to overcome them. The tool and its underlying method should be capable of accepting incomplete and inconsistent information about coordination. EPM solves this by using the dually interlinked intermediate result network and by accepting rough plans and schedules by help of effort considerations only. EPM also supports the decentral resource management what is lacked by PMS on the marketplace. EPM´s multi-user ability overcomes the problem of central data input and output. Decentral resource management, the permission of inconsistencies and incompleteness together with the multi-user ability reduce the amount of effort to service the project data to a minimum and distribute the rest of effort to the whole project team. And as shown EPM supports by using the decentral resource management approach intrinsically multiproject management. Very important is that EPM has a very strong measure for project goal attainment which directly relates to the construction of customer-supplier chains. This is also supported by establishing a reporting and documenting system. Last but not least EPM supports the evaluation of project experiences in the audit phase. Therefore we conclude that the project management method and EPM are appropriate for the management of Concurrent/Simultaneous Engineering product development processes in general. And what do the customers say about EPM? Because there was not yet enough time to test EPM in a normal product development process at DASA and TEMIC experiences with a basic version of it should be mentioned. This version is in operational use at LOEWE, a medium sized company in the consumer electronics branch which has already implemented teams since 1992. This test proved the major direction of the concepts to be feasible. Prototypes of EPM are right now installed at DASA, TEMIC, Hidrosorefame and ASF, Berlin. Further results will be available in the near future.

The proposed project management also fulfills the requirements to business process management methods. As shown before the process orientation is inherent to Concurrent/Simultaneous Engineering. EPM itself can be analysed as a system which supports the initiation, arrangement and execution transactions (Ferstl/Mannmeusel, 1995) which have to be done during a project to build up, coordinate and realise a specific product development project. Formal aims are defined by an executive level. And there is the negotiation process for planning and conflict resolution during the runtime of the project in which the systematic involvement of the client into the process design is assured (comp. Hess et.al., 1995) The processes which are combined on the project level are subject to customer orientation (Hess et.al., 1995). Every sub-process is defined by its intermediate-results which are negotiated with the client who relies on it. The client can refuse the result, and after finishing a project the definition of the intermediate-results can be reviewed and changed. Business process management is an evolutionary approach (Deiters et.al., 1995; Scheer et.al., 1995). The method underlying EPM supports the continuous improvement process required by Hess et.al. (1995). Every single product development process is planned and scheduled on the basis of the experiences of former projects. Milestone plans, workpackage lists defining customer-supplier relations are updated and improved after finishing every single project. In the case of conflict during the runtime of projects the EPM-method supports the decision making process which in the end lead to an updated plan. Changes in the process during its runtime are communicated to all teams. EPM supports the learning cycle by storing data and documents about the actual product development process. Their analysis may lead to improvements. Summing up, EPM and its underlying method are first steps to realise supporting tools to apply business process management methods to open and creative processes. The experiences with the recently implemented prototype give hope to continue to try to support the here proposed planning and coordination method.

5 Debates and Developments

EPM has been developed in the debate about adapting project management methods and tools to Concurrent/Simultaneous Engineering. But in this realm there are further concepts and approaches which could be incorporated in business process management for open and creative processes, too. Eversheim et.al. (1995) propose a product-neutral development plan which should be the basis for project management. This neutral plan is the result of a benchmark driven optimisation. The plan defines the logical process sequence, result exchanges and consists of review and release events. Obviously, this approach could be very well incorporated into the EPM method as long as EPM does not support the process limitation and makes no statements about initial milestone and workpackage structures. Krause, Golm (1995) follow a different approach. Their aim is to combine a product model with an process model, a resource and organisation model. The status of the product model is changed by activities. Therefore the information which is produced in the process model must be represented in the product model. And in order to define the processes which change product data they must be defined with regard to the product model. All models are optimally combined by an optimisation algorithm depending on the project type which is described by its degree of innovation, task concentration, etc. Wildemann (1994) wants to apply the Just-in-Time (JIT) concept to the management of product development processes. His approach can be interpreted as standardisation, as increasing the degree of deterministic processes. His concept could be implemented on the subprocess level and for projects which have a low degree of innovation. Obviously, these different approaches do not exclude each other. They could be combined, they should be implemented under different circumstances in a different manner. The important task to find out which of the proposed tools and methods should be implemented in which specific case remains. There are some results from Concurrent/Simultaneous Engineering research in which process types and methods to define the development processes are presented (Specht/Beckmann, 1996; Wheelwright/Clark, 1992). But further research is necessary here.

EPM is now available in a prototype and expanding ideas are already there. EPM does not plan. Even in very complex development projects with many parallel and interdependent tasks it would be helpful to support the elaboration and selection of alternative schedules. In the successor of EPM which is called TOPP (team oriented project planning system) (Bullinger et.al., 1996) such a component will be realised. It bases upon a constraint propagation approach which identifies all true schedules for a time frame which is defined by a consistency frame. The identified schedules will be ranked according to important process quality describing variables. This will help to focus the discussions between the teams and the project manager. Further improvements regard the workflow component. Project management in itself can be understood as a set of business processes. Many of them could be standardised such as input of controlling information or result exchanges, other processes like negotiations could be framed by supporting them with a documentation management module. Another idea to structure negotiation processes is to apply the speech act theory to them (Winograd, 1987; Austin, 1962; Searle, 1969). Last but not least there are ideas whether to incorporate a quasi-knowledge-based planning component based on intelligent mechanisms which evaluate given experiences from recent projects. This intelligent mechanisms do only propose conclusions - to make them is the task of the users. This maintains the autonomy of the user (Bullinger, Kornwachs, 1990).

Some remarks at the end. We are very interested in the question how the teams in different enterprises will make use of the offer EPM. Will they have in the end very stable intermediate result networks? Will they tend to work with defined milestones? How will they organise their conflict and inconsistency management and how relate all this questions to different types of complex product development processes?

References

Austin, J.L. (1962) *How to Do Things with Words*. Cambridge, MA.
Bergner, J. (1996) TEMIC TELEFUNKEN microelectronic GmbH, in *Concurrent Simultaneous Engineering Systems: The way to successful product development* (eds. H.J. Bullinger, J. Warschat), Springer, London.
Berndes, S., Stanke, A. (1996) A Concept for Revitalisation of Product Development, in *Concurrent Simultaneous Engineering Systems: The way to successful product development* (eds. H.J. Bullinger, J. Warschat), Springer, London.

Bürgel, H.D., Kunkowsky, H.R. (1989) EDV-unterstütztes FuE-Controlling. *Technologie & Management*, 3, 25 - 32.

Bullinger, H.J., Kornwachs, K. (1990) *Expertensysteme*. Beck, München.

Bullinger, H.J., Warschat, J, Berndes, S, Stanke, A. (1995) Simultaneous Engineering, in *Handbuch Technologiemanagement* (ed. E. Zahn) Schäffer-Poeschel, Stuttgart.

Bullinger, H.J., Warschat, J, Wörner, K., Wißler, K.F. (1996) Rapid Product Development: Ein iterativer Lösungsansatz zur Verkürzung der Entwicklungszeiten in dynamischen Umfeldern. *VDI-Z*, 5, 38 - 41.

Clark, K.B., Fujimoto, T. (1992) *Automobilentwicklung mit System: Strategie, Organisation und Management in Europa, Japan und USA*. Frankfurt/Main, New York.

Davenport, T.H. (1993) *Process Innovation: Reengineering work through information technology*. Ernst & Young, Boston.

Deiters, W., Gruhn, V., Striemer, R. (1995) Der FUNSOFT-Ansatz zum integrierten Geschäftsprozeßmanagement. *Wirtschaftsinformatik*, 5, 459 - 466.

Deiters, W., Striemer, R. (1994) Workflow-Management: Chancen und Perspektiven prozeßorientierter Workgroup-Computing-Systeme. *DV-Management*, 3, 99 - 104.

Domschke, W., Scholl, A., Voß, S. (1993) *Produktionsplanung: Ablauforganisatorische Aspekte*. Springer, Berlin.

Dworatschek, S., Hayek, A. (1990) *Marktspiegel Projektmanagement Software: Kriterienkatalog und Leistungsprofile*. Verlag TÜV Rheinland, Düsseldorf.

Eversheim, W., Marczinski, G., Lauffenberg, L. (1995) *Simultaneous Engineering*. Springer, Berlin.

Ferstl, O.K., Mannmeusel, T. (1995) Gestaltung industrieller Geschäftsprozesse. *Wirtschaftsinformatik*, 5, 446 - 458.

Ferstl, O.K., Sinz, E.J. (1993) Geschäftsprozeßmodellierung. *Wirtschaftsinformatik*, 6, 589 - 592.

Galler, J., Scheer, A.W. (1995) Workflow-Projekte: Vom Geschäftsprozeß zur unternehmensspezifischen Workflow-Anwendung. *IM Information Management*, 1, 20 - 27.

Hammer, M., Champy, J. (1995) *Business Reengineering: Die Radikalkur für das Unternehmen*. Campus, Frankfurt/Main, New York, 5. ed.

Hess, T.-H., Brecht, L., Österle, H. (1996) Stand und Defizite der Methoden des Business Process Redesign. *Wirtschaftsinformatik*, 5, 480 - 486.

Kessler, S. (1996) SIFRAME: The CONSENS Framework, in *Concurrent Simultaneous Engineering Systems: The way to successful product development* (eds. H.J. Bullinger, J. Warschat), Springer, London.

Krause, F.L., Golm, F. (1995) Kennzahlengetriebene Optimierung von Entwicklungsprozessen. *ZWF-CIM*, 7 - 8, 372 - 375.

Lyotard, J.F. (1993) *Das postmoderne Wissen: Ein Bericht*. Passagen, Wien, 2nd. ed.

Malik, F. (1986) *Strategie des Managements komplexer Systeme*. Bern, Stuttgart, Haupt.

McGrath, M.E., Anthony, M.T., Shapiro, M.R. (1992) *Product development success through product and cycle time excellence*. Butterworth Heinemann, Stoneham, MA.

Miotti, L. (1996) IMS: Information Management System, in *Concurrent Simultaneous Engineering Systems: The way to successful product development* (eds. H.J. Bullinger, J. Warschat), Springer, London.

Möhrle, M.G. (1991) *Informationssysteme in der betrieblichen Forschung und Entwicklung*. Verlag H. Schäfer, Bad Homburg.

Nippa, M. (1995) Anforderungen an das Management prozeßorientierter Unternehmen, in *Prozeßmanagement und Reengineering: Die Praxis im deutschsprachigen Raum* (ed. Nippa, M., Picot, A.) Campus, Frankfurt/Main, New York.

Patterson, M.L. (1993) *Accelerating Innovation: Improving the Process of Product Development.* Van Nostrand Reinhold, New York.

Picot, A., Franck, E. (1995) Prozeßorganisation: Eine Bewertung der neuen Ansätze aus Sicht der Organisationslehre, in *Prozeßmanagement und Reengineering: Die Praxis im deutschsprachigen Raum* (ed. Nippa, M., Picot, A.) Campus, Frankfurt/Main, New York.

Poser, H. (1990) Wissen und Können: Zur Geschichte und Problematik des Wissenschaftstransfers, in *Handbuch des Wissenschaftstransfers* (ed. Schuster, H.J.), Springer, Berlin.

Scheer, A.W., Nüttgens, M., Zimmermann, V. (1995) Rahmenkonzept für ein integriertes Geschäftsprozeßmanagement. *Wirtschaftsinformatik,* 5, 426 - 434.

Searle, J.R. (1969) *Speech Acts.* London.

Sinz, E.J. (1995) Editorial zum Schwerpunktthema Geschäftsprozeßmodellierung. *Wirtschaftsinformatik,* 5, 425.

Specht, G, Beckmann, C. (1996) *F&E-Management.* Schäffer-Poeschel, Stuttgart.

Stalk, G., Hout, T.M. (1990) *Zeitwettbewerb: Schnelligkeit entscheidet auf den Märkten der Zukunft.* Frankfurt/Main, New York.

Schwarze, J. (1990) *Netzplantechnik: Eine Einführung in das Projektmanagement.* Herne, Berlin, 6. ed.

Vilsmeier, J. (1996) Dasa Military Aircraft Division, in *Concurrent Simultaneous Engineering Systems: The way to successful product development* (eds. H.J. Bullinger, J. Warschat), Springer, London.

Wheelwright, S.C., Clark, K.B. (1992) Creating Project Plans to focus Product Development. *Harvard Business Review,* 2, 70 - 82.

Wildemann, H. (1994) Organisation und Projektabwicklung für das Just-in-Time Konzept in F&E und Konstruktion (Teil2). *zfo,* 2, 128 - 133.

Distributed Co-operative Modelling of Production Systems

Bernd Scholz-Reiter and Dörte Bastian

Chair of Industrial Information Systems, Brandenburg Technical University of Cottbus, P.O. Box 101344, D - 03013 Cottbus, Germany, {bsr, bastian}@iit.tu-cottbus.de

Abstract. For some years the understanding of planners has been rising which has meant that not only the manufacturing and information technology of an enterprise are considered important factors for success but also its staff, organisation and processes, particularly the mutual connections between these components. Various specialists should also collaborate in order to complete such demanding, complex and extensive tasks. Thus it is necessary for experts from various fields such as factory planners, business organisers, and systems analysts to be involved in planning manufacturing. Much distributed planning consists of tasks within groupware processes. By means of the computer-aided concept presented here, the co-operative and distributed collaboration of planning experts is considerably improved and developed with regard to the efficiency as well as the quality of planning projects.

Keywords. Business Process Modelling, Manufacturing Simulation, Distributed Co-operation, Integration of Data, Object-oriented Metamodel, Groupware

1 Missing connection between distributed modelling tasks

Planning is supported by *various models and modelling techniques* as well as their implementation using computer-aided tools in order to take account of the variety of planning information and requirements for the modelling process.

Manufacturing simulators, business process modelling tools and CASE-tools used by respective experts are available for designing an improved model of the firm. Planners in different departments in a firm use these tools to complete work, the tools being implemented on distinct platforms physically distributed and either parallel or asynchronous with regard to time. As a consequence the process of modelling can be characterised as distributed multi-platform-modelling.

Until now there has been no connection between such application systems for modelling either in a logic-conceptual or physical sense.

This *methodological lack* results in much activity concerning coordination between systems and also the preservation of consistent partial models.

Influences on and interactions between determinants in disparate models are not deliberated. *Lacking technological facilities* for connections between modelling tools cause a large number of media breakes and also prevent a fast exchange of information.

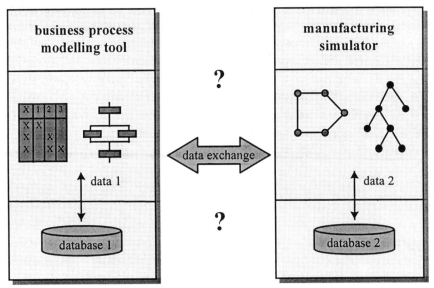

Fig. 1: Distribution of modelling tools

A methodological combination of and hardware-based connection between modelling tools is necessary for carrying out all planning tasks efficiently and with the help of a computer e.g. for the exchange of documents or planning models. By means of significant facilities which make the combination of distributed application tools possible, the essential improvement of information exchange between planners in a firm can be observed. These improvements mean that information can be transferred with a greatly reduced transmission time in comparison with traditional means of transferring information (e.g. by post) and provide ways of transmitting information other than by telephone (for instance using graphics or planning models).

In addition, the distribution of planning tasks enables a parallel accomplishment of different processes in modelling projects and as a result a reduction in modelling time.

2 Integration of data as a basis for the methodological combination of modelling tools

Some computer-aided tools can be used for disparate problems while planning business processes and manufacturing structures in a firm. These tools enable a transparent representation of relevant planning subjects and their various relationships in the field of manufacturing. *Manufacturing simulators* are used for designing and improving manufacturing systems and material-related processes. *Business process modelling tools* are applied in order to design and develop information systems and communication structures in the production department.

These modelling tools function on various logical data models and physical databases due to a number of terminologies and the development of diverse systems on different levels of hard- and software, by a variety of software specialists. Despite the fact that they describe the same subjects in a firm, there are neither connections to combine the tools nor ways of transmitting current planning models.

In other words *separated data management* is a major obstacle to flexible collaboration between enterprise planners during computer-aided modelling and the simulation and analysis of manufacturing processes.

With this *unsatisfactory situation* as a basis the individual divisions of modelling are methodologically integrated and then implemented in a computer-based modelling environment to support simultaneous modelling.

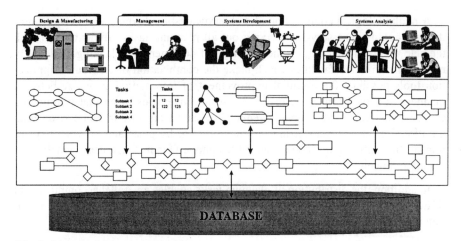

Fig. 2: Integrated data management

A *common logical data model* on which several tools can work was necessary in order to exchange all modelling data between application systems. This uniform common data model was designed through a process of data integration which required drafting an integrated schema on a logical level.

An object-oriented metamodel of manufacturing was developed by identifying basic objects in the integrated schema and specifying relationships between them. Thus the description of all occurring data objects is guaranteed, particularly with regard to the preservation of semantics for exchanging information. Logical connections between data are directly laid aside within the database through which a better representation of the user world is obtained. In addition such an integrated *object-oriented metamodel* contains all modelling components of its modelling sources and is a conceptual basis for a common, physical object database.

In this object model, objects with the same characteristics are encapsulated into classes of objects as is done in reality. Notions of departments, their concepts and ways of thinking or abstractions of an application field, determine objects by their attributes and methods as well as by their structure of classes.

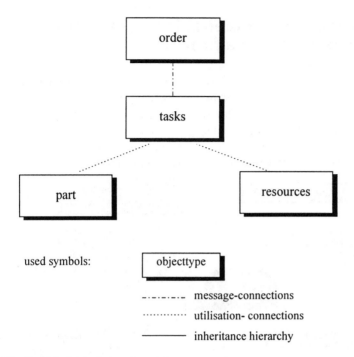

Fig. 3: Object-oriented metamodel

Specialised terms are determined by the definition of object classes
- order,
- task,
- part,
- resources

and their interrelationships.

Object orientation enables complex business processes to be modelled in a flexible manner and supports the improvement of manufacturing systems and information and communication structures in the field of production.

Based on the principle of inheritance, the object-oriented concept also enables the enlargement and reuse of object-oriented models so that definitions of new objects can be constructed from existing classes. Thus *integration of future systems* of the different tool classes for modelling manufacturing systems and information and communication structures is guaranteed.

Besides domain-specific terms, the object model contains additional information on versions and configurations as well as specific knowledge regarding for example planning data and project information.

3 Importance of computer-aided collaboration between planners in a firm

The realisation of planning tasks on the shop floor is initiated by forming work-groups consisting of employees involved in manufacturing planning and organisation as well as specialists from other relevant areas. After the planning stages have been assigned to different members of the group, the subtasks are worked on sequentially, but also simultaneously. Apart from these pure modelling tasks, there are a number of groupwork processes as the diagram below illustrates.

Fig. 4: Groupwork processes

Distributed modelling tasks are carried out *by means of existing tools*. The benefit of using specific and highly sophisticated tools is that they do not have to be re-invented, but can be used in the same form in which they are available on the market. In addition, planners in various fields can continue working with tools they are familiar with and which were developed for their individual tasks. Planning experts are not stretched by the mega-functionality of a comprehensive tool.

The complex modelling tasks are performed not only in a physically distributed manner, but also co-operatively which means *in collaboration with* all other *involved planners*. There are many interdependencies between domain-specific tasks of the planners, which must be considered in the planning process. The distribution of tools alone is not sufficient for improving collaboration between planning experts.

Furthermore, in order to take account of the many-sided dependencies between the subtasks of planning, individual modelling specialists spend much time involved in other *groupwork processes* such as meetings, discussions and decision-making with regard to variations to the planning process and resolving conflicts.

Presently the primary means of communication between planners during groupwork and during the coordination of subtasks when designing the shop floor are those traditional means such as by internal post, by telephone, using circulars and through individual arrangements (see Fig. 5).

Thus *conventional ways of working* use no computer support, but instead work is carried out through oral or written agreements between and requests from co-workers in the planning group. The result of this is that many working steps and measures of the shop floor e.g. KANBAN-quantity were not considered in terms of their mutual connections to other business factors during the planning process and thus had to be revised. These modifications often had certain inadequacies, which required further planning.

Face to Face

By phone

By mail

Group discussion

Fig. 5: Traditional means

Disadvantages concern the following traditional means of supporting co-operative, distributed modelling processes:
- a great number of media breaks
- multiple inputs during modelling
- working with partially old planning scores
- inconsistencies in isolated planning models and as a result high expenditure due to preservation consistency
- decreased quality of planning through information deficits
- great organisational expenditure and costs in order to arrange an appointment for instance with regard to a conference
- delays to planning work e.g. by sending planning models by post.

The increasing complexity and interdependencies in manufacturing make *manual planning* by specialists working co-operatively in groups *extremely difficult and hard to organise* which results in high costs and poor results. Thus an improved support is necessary for distributed collaboration between firm planners during the targeted solving of planning problems on the shop floor. Until now the possibili-

154

ties presented by co-operative, distributed problem solving have not been used for the computer-aided modelling of manufacturing systems.

4 Concept of distributed, co-operative modelling

Because of the above-mentioned reasons, the design of a computer-based toolbox for distributed, co-operative modelling was necessary in order to efficient support the collaboration between various firm planners, who work with different model-ling tools.

4.1 Structure of the toolbox

Apart from separately modelled tools, the toolbox consists mainly of three compo-nents.

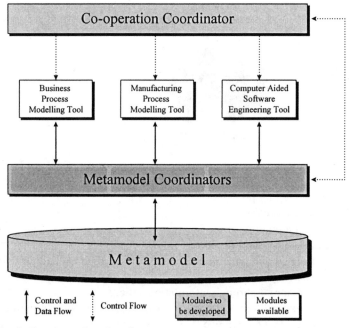

Fig. 6: Groupware-based toolbox

The basis and integration kernel of the toolbox is a comprehensive, *object-oriented metamodel* of manufacturing which enables all components and their interdependencies to be stored dynamically and reproduced from different points of views.

The *metamodel coordinators* are responsible for converting all tool-specific data which is connected to data of other tools to and from the common metamodel. First the metamodel coordinators obtained the level of converters. In a second step they were equipped with some kind of intelligence for processing the multi-directional transformation.

The process of simultaneous modelling is concerned with both information-related and material-related aspects and was implemented by a component governing all tools - the *co-operation coordinator*. This controls the modelling process of co-operators working in groups through a higher system for distributed, co-operative problem solving using groupware concepts. The development of the co-operation coordinator was closely connected with design of the modelling methodology and the graphical user interface.

4.2 Analysis of co-operative problem solving processes

During the groupwork process of distributed, co-operative modelling various types of collaboration can be differentiated with regard to different forms of co-operation. Some of these co-operative types of work for designing manufacturing systems by a mostly physically separated group of planners needed more intensive support by computers than others.

According to Johansen's division of co-operative activities concerning group-work, the dimensions time and space can be differentiated. In contrast Sogard differentiates between time and focus of work. These two matrices can be summarised and then displayed as follows (see Fig. 7).

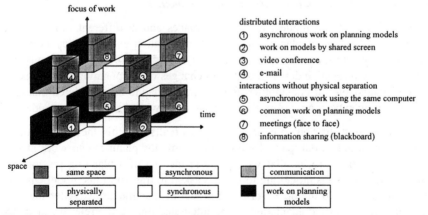

distributed interactions
① asynchronous work on planning models
② work on models by shared screen
③ video conference
④ e-mail
interactions without physical separation
⑤ asynchronous work using the same computer
⑥ common work on planning models
⑦ meetings (face to face)
⑧ information sharing (blackboard)

Fig. 7: Co-operative work types of planning

In the dimension of time it is differentiated between synchronous and asynchronous activities. Because of spatial distance between the planners, they can work either in direct co-operation or physically separated. The dimension focus describes the kind of interactions differentiating between communication or working on a planning model. The specifics of a computer-based realisation of the various kinds of collaboration are explained in the following chapters.

4.2.1 Distributed interactions

Especially the collaboration based on distribution needed particular technical means like connection of computers and special software with groupware functionality. The communicational bases of distributed, co-operative planning are explained in one of the next chapters.

• *Asynchronous, distributed communication*

Because of the physical separation of the planners, transmission capabilities for texts, pictures and planning models were accomplished through multimedial know-how. This multimedial kind of interaction with the computer is more efficient and user-friendly than in- and output of texts only. Support of explicit, computer-based communication between firm planners at different times is currently possible through a large number of e-mail systems. Explicit communication is based on sending and receiving messages by means of a communication channel. As a result, information sharing between different firm planners is realised and, in addition, the number of media breaks was essentially decreased and was kept as small as possible.

Electronical message transmission was achieved by using the groupware system Lotus Notes within the toolbox for the support of distributed, co-operative modelling.

• *Asynchronous, distributed work on planning models*

Designing planning models in different places and at different times needed flexible coordination of planning tasks and information exchange depending on the modelling activities. The specific problem of this type of interaction consisted in making the respective developments and changes of the planning model visible and understandable for one another. Besides this, it had to be considered that only those planning experts are informed who have to do with changes of the model.

Asynchronous, distributed collaboration was realised within the co-operation - coordinator, which is one part of the toolbox. Now the planners can work at first with the tools they are familiar with. Only at the end of a planning session, i.e. when a user closes his modelling tool, he is asked to transform his model changes or modifications into the object-oriented model of manufacturing. After the confirmation of the desired adaptation of the integrated model, the relevant metamodel-coordinators are automatically started. These transformation routines convert the modelling data into the model of manufacturing and store the extensions or changes made in the object-oriented database.

In a second step, converters were further developed with regard to a sensible combination of modelling tools in order to realise an asynchronous design of the planning models. So they considerably go beyond a simple interface for exchanging data. Now they send the consequences of the model changes electronically to all planners involved. Reasons for information can be the modelling components which were deleted, newly defined or modified in some parameters. Before modelling data are sent, they are prepared considering the modelling possibilities of the tools according to the view of the respective planners

- *Synchronous, distributed communication*

Communication between physically separated members of a planning group at the same time, for instance in order to carry through a video conference, needs very intensive technical support like audio and video connections between offices which are especially equipped or between rooms with cameras, microphones and monitors etc. Technical support by means of video conferences is only justified, when collaborators work in very different places, so that they cannot meet personally because of costs or time.

The expenditure required for this type of teamwork for single-item and small batch producers is disputable. Thus within the framework of this project computer-based support of synchronous, distributed communication is presently not tried.

- *Synchronous, distributed work on the planning model*

Common work on problem solutions during planning is facilitated and supported by using computer technology. The expenses required were primarily due to the necessary software, which shall constitute the basis for a collective editing of planning models. Collaboration between several planning experts is essentially improved by using a common monitor image (shared screen). Changes made by one planner become at once visible to all other collaborating planners. So group members work out their opinions and suggestions directly at the planning model and, in addition, they immediately come to an agreement about changes and modifications. The common work on a planning problem by means of a shared screen leads first of all to a reduction of planning time during the design of a model and regarding the agreement on the modelling solutions.

4.2.2 Interactions without physical separation

Co-operation between planners in the same place needed far less support than distributed interactions. Here mainly time dimensions had to be overcome, i.e. to garantee access to common information and data stocks (information sharing). Planning experts should be able to use and handle information which is important for their work even when the other co-workers are absent.

- *Synchronous communication*

If planners work at the same time and in the same place, it is mostly consultations, work sessions or conferences. For the single-item and small batch producer investigated the expenditure to prepare these meetings, which contain e.g. the creation of a schedule, is rather low. In addition, the costs to carry out such meetings are mostly not so high, because only few planners are usually involved in the planning process. Discussions and meetings are relatively well-structured and orderly. These tasks are not implemented in the toolbox, because in comparison with traditionally supported meetings a significant improvement of meetings by means of computers would not be expected here.

In order to support the coordination of time, a group-schedule programme was introduced to accompany the planning process. Here a considerable time saving and decrease in the organisational expenditure are observed in contrast to ordinary agreements mostly made by telephone. In addition, by using the computer-based group-scheduler, it is guaranteed to recall an important deadline or to remember necessary documents for a meeting.

- *Synchronous work on the planning model*

Planners working together in the same place on crucial points in planning did not need any special help. Here different planners collaborate directly at one computer in order to work out a model of the manufacturing system. But planners should allways check if distributed collaboration is more efficient. During this, all changes made are immediately considered in the modelling tool and are displayed immediately to all planners involved. Moreover, every planner can make his own comments and suggestions, whereas planners can only sequentially make changes or modifications at a personal meeting. During distributed collaboration, it could be possible that more planners are involved in the decision-making process, leading in a personal session to problems with regard to place in front of the computer used.

- *Asynchronous communication / Asynchronous work on the planning model*

If co-workers want to co-operate at different times and in different places, until now a blackboard with information on it concerning the coordination of work tasks as well as subject matters of planning work has usually been used. Thus conventional information exchange between planners and announcement of new data mean that every planner informs himself about new deadlines and project data on the blackboard. In future collaboration will here be primarily supported by computer-based information sharing. A better result concerning quality would be reached by sharing the information, because every co-worker has the necessary information at his disposal at all time. For this purpose, a blackboard was implemented into the toolbox on which the information available for all planners can be acquired.

4.2.3 A special form of collaboration - simultaneous work

In addition to the forms of co-operation described above, a parallelisation of various processes of modelling is possible through the distribution of planning tasks which results in reduced modelling time.

However parallel work on objects is not possible using conventional application systems such as relational databases. Instead, the objects are blocked against further access during the working whole time (transaction), that means they are exclusively handled. But some problem processing takes a lot of time, so that finding a solution can last several days or weeks. Therefore it is not sensible to block these objects against exclusive access during their handling. Thus parallel processing and changing of common components in the integrated model of manufacturing by using different modelling tools needs particular technological help.

The co-operative use of common data in a planning group whose members work closely together can only be supported by means of a new processing model. In addition to mechanisms for competing coordination (e.g. lock-outs; atomic, isolated courses), an extension of the ACID-transaction model for the handling of database access which takes a long time and is concurrent is required. In order to manage simultaneous collaboration, appropriate coordination mechanisms must be used which cause a system-controlled decline of exclusive data access. Therefore mechanisms such as messages, temporary integrity conditions and negotiations which can partly be automated are required in order to support co-operative coordination.

If the processing of modelling objects is completed during long lasting, concurrent transactions, incompatibilities and conflicts with planning objects modelled parallel can occur. Demand for coordination and co-operation is created through changes to the common data stock during simultaneous modelling. In other words, whenever parallel access to common objects of the planning model is gained, agreement of interactions between planners is indispensable. Only this way the common planning task can be completed without any problems concerning the design of manufacturing and the maintenance of consistency of the integrated model.

4.3 Project-accompanying functions

Besides mechanisms which are directly focused on execution and improvement of co-operative modelling, overall-modelling functionalities must be developed and implemented in the toolbox. These modelling-accompanying concepts are directed at the facilitation of groupwork which goes beyond pure support of continuous modelling.

4.3.1 Definition of project information in the object model

Information about the structure of the project group and the tools used must be announced and saved at the beginning of a planning project. Therefore a project modul was developed which establishes all necessary data based on user dialogues when a new project is created. Groups can be produced, tools can be allocated to group members and project information can be looked at by using this project modul.

The necessary information about project groups and group members, about the modelling tools used and their dependencies are stored in objects of the integrated model of manufacturing. For this purpose the following object classes together with their attributes and methods as well as some abstract classes were worked out and implemented

- Team
- Team member
- Tool.

During the planning project, the project information stored in the object model is used to coordinate the interdependent subtasks of different co-workers. This coordination between separate tasks is part of the perfomance spectrum of project management. It is supported within the toolbox by means of suitable developed software-moduls. On the one hand, computer-based execution of the project management results in benefits regarding efficient coordination of subtasks during planning and, on the other hand, relieves the project leader and the team members of necessary coordination tasks.

4.3.2 Project-spread group-scheduler

The toolbox requires a function accompanying planning work which can be used to simplify and speed up making appointments within the planning group in comparison with conventional means (e.g. phone, circular letters, mail). This groupware tool for time and resource management is the so-called group scheduler. With its help, from now on it will be possible to make common appointments for group discussions, sessions etc. with considerably reduced organisational expenditure.

4.3.3 Further planning-accompanying functions

Additional tasks which must be carried out through the whole lifetime of a project affect the following functions:
- progress control of the project
- storing the modelling history
- integration of new tools
- designing the user interface.

These tasks are performed by means of a computer since this way an increase in efficiency of collaboration as well as a relief of the project leader is achived.

4.4 Basis of flexible collaboration

During the co-operative accomplishment of a planning project, the individual interdependent working steps mainly concern planning and modelling tasks. These problem definitions belong to those tasks which are very difficult to structure and are characterised by
- considerable difficulties when structuring the tasks
- indefiniteness of required information
- difficulty in predicting the order of subtasks
- uncertainty of finding a solution
in comparison with which can be easily subdivided.

Thus the individual working steps within distributed, co-operative modelling (which are dependent on one another) could not be planned, controlled, observed and executed by means of predetermined flow management. In contrast to well-structured tasks they are synchronised on dynamically specified access points. Until now these problem definitions have not been supported by computer technology.

Due to the fact that the different and comprehensive planning tasks of a planning project are not described in terms of a structured, well-defined business process „planning", flexible possibilities had to be applied in order to synchronise, mutually influence and manipulate the interdependent subtasks of planning problems. In contrast to a given workflow schema, the coordination of work tasks within modelling is based on a very large number of communication processes. As a consequence, flexible teamwork had to be controlled by means of a communication-based coordination model. Synchronisation of more unstructured modelling processes was achieved by
- dynamic identification of demand for coordination
- event control
- a large number of communication processes.

A schematic sketch shows the sequence of necessary steps to coordinate the interdependent modelling tasks (see Fig. 8).

If the metamodel-coordinators identify changes during the transformation of modelling data into the object model, the planners concerned are informed. For that the changed modelling data are firstly prepared in accordance with the respective view of a user and then sent automatically to the planners through the metamodel-coordinator using a communication modul. Thus planners are actively required to include changes in the common data stock.

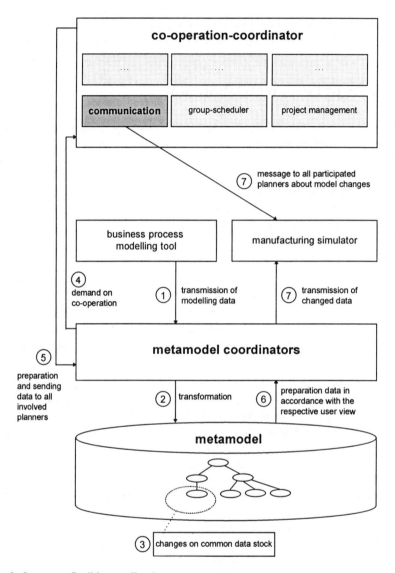

Fig. 8: Steps to a flexible coordination

4.4.1 Realisation of dynamic coordination between action plans

The coordination between separated modelling phases is now flexibly and dynamically realised by the system-spread co-operation modul. In the following part essential work and implementation steps are described.

- *Event control*

For a flexible coordination of subtasks an extension of the modelling tools, which up to now have been isolated, was necessary with regard to a system-specific co-operation function. On the one hand, this function causes events concerning the system condition to be announced. On the other hand, this function helps to perform an action which results from an event of a tool. In order to implement this tool extension, a C++-routine was developed which sends a message when a modelling tool is closed. This message causes the suitable metamodel-coordinator to be called up automatically.

- Dynamic identification of demand for coordination

The identification of demand for coordination within the metamodel-coordinators is dynamically noted by locating changes in the integrated metamo-del. Thus only such changes are taken into account which have an effect on the planners involved.

Company objects could be
- newly defined
- erased or
- modified
through changes in the object-oriented model.

Identifying changes is realised by methods belonging to the relevant object classes of the object model of manufacturing with the help of query requests. Existential, single element or usual queries are used depending on their purpose. The relevant collections which store the model objects are searched through for changes by using queries for which some search criteria must be declared.

Newly defined objects

When in the respective object collection no object has been found with the relevant object ID, a company object is newly defined. Existential queries which allow quick processing of the request are used for this kind of search. If the metamodel-coordinator finds out that an object has been newly defined, this object together with all its fixed object parameters is then sent to the planners concerned.

Modified objects

If the searched object has been found out by an existential query, it is necessary to check all object parameters for newly defined or modified values. For that the object desired is firstly picked out of the collection using a single element query and then examined in detail. Objects in which modifications have been found are sent to the participating planners with a comment regarding the parameters of model elements concerned.

Erased objects

Erased objects are determined at the end of a data transformation by comparing the file dates. In contrast to newly defined or modified objects they can not be identified explicitly by metamodel-coordinators using queries regarding collections. Instead it is necessary to examine the objects with regard to the current dates of the files from which the data are converted into the object model. Those objects in the collection which contain an old date were obviously erased during the last modelling process.

* *Communication processes*

The communication modul ensures that the planners who are connected with model changes are informed and receive all relevant data according to their user view. For that the e-mail functions of the communication modul which realise a message exchange are automatically called up. Thus the action plans of the individual project members are dynamically coordinated taking the interdependent influences of the company´s elements into account.

4.5 Communication - basis for teleco-operative modelling

In addition to usual e-mail functions, the communication modul supports the quicker and more efficient performance of planning tasks. It adds to the electronic sending of messages the possibility of mailing further information to the relevant planners such as planning models and documents including model changes. This multimedial kind of communication is more user friendly than the transmission of texts only.

* *Aspects of technical realisation*

A local area network (LAN) is the basis for quicker communication connections between planners. The connection between the individual computers is realised through a Windows NT-network based on TCP/IP-protocols. The software of the server runs under Windows NT, the client-software under Windows 3.11, Windows 95 as well as Windows NT. Lotus Notes is used as communication and groupware environment which supports a large number of platforms, network systems and protocols.

A special connection between the respective application system and Lotus Notes had to be additionally created for the electronic data exchange of model modifications as well as for the transmission of project information into the object model. For that aim, Lotus Notes provides both a DDE (Dynamic Data Exchange) and an API interface, whereas in the course of this project the DDE functionalities are used. For the use of these functions it was necessary to create and register a DDE server which realises the connection buildup and cleardown as well as the data transmission between some Notes Clients and various application systems.

- *Communication between planning experts*

The communication between planners is carried out on the basis of e-mails where messages can be sent to individual people, groups of different people or the whole project team. Every user must have a Lotus Notes client in order to receive or send e-mails.

- *Notification of planners concerned*

As mentioned above Lotus Notes provides a DDE interface which enables the information technical connection between all planners involved in the planning project in order to transmit messages to and from different modelling tools. Sending, receiving and also storing of messages is realised by using both Lotus Notes and a communication modul developed during this project.

Access to the respective Notes client for the purpose of transmitting model changes is established by its DDE interface. This interface is called up by a communication modul, which was especially developed for this, and encapsulates the different essential DDE functions on the basis of a DDEML manager (Dynamic Data Exchange Management Libary). This means in other words, the mail functions were implemented as an interface which, if required, is called up automatically by the metamodel-coordinators and gets access to the relevant Notes client through DDE (see Fig. 9).

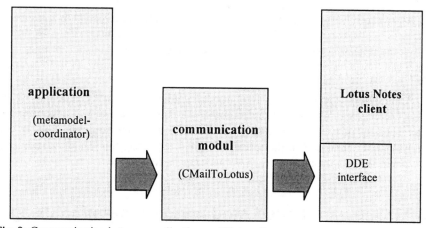

Fig. 9: Communication between application and Notes client

Failures during the initialisation of the DDE interface as well as during the establishment of a connection to Lotus Notes are recognised and trigger off an exceptional procedure. As a result, the user is given detailed information about the failure and its reasons.

166

- *Transmission of project and planning information from Lotus Notes into the object model*

Definitions regarding the structure of the project team and the functionality of planning tools as well as the assignment of the modelling tools used are filed as project and planning data in the objects of the tool-spreading metamodel. This information has been gained using Lotus Notes. For that, a connection between the different Notes clients and the object-oriented database was required (see Fig. 10).

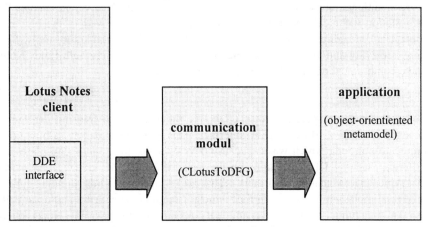

Fig. 10: Communication between Notes client and application

Access to other applications (e.g. the object-oriented database) was implemented with a callback routine. The communication modul guarantees the assignment between the DDE interface of a Notes client and the suitable callback routine.

- *Warning and memory function*

Additionally, the communication modul contains a warning function which does not only signal new mailings to users, but also the reception of new information because of modified planning models. Occuring transmission failures are immediately announced to the users concerned. By using this functionality users are reminded of appointments such as meetings or conferences.

5 Groupware-based modelling of manufacturing systems

A groupware-based *concept for distributed, co-operative problem solving* was developed in order to effectively support collaboration during planning projects. Therefore groupware concepts from computer science were adapted to the specific demands of modelling manufacturing systems and then implemented into a computer-aided planning component.

5.1 Groupware moduls within the Co-operation-Coordinator

The Co-operation-Coordinator - as the controlling component of the toolbox - consists of the following workgroup approaches for the realisation and execution of overall modelling in addition to system-specific functions which are related to the modelling process:

- sending electronic mails
- initiating data exchange between modelling tools and the metamodel
- flexible coordination of task flows
- arrangement of common appointments
- electronic information exchange (information sharing)
- functions of project management
- controlling the common work on documents.

Groupware-systems aim at providing common access to resources for several co-workers of a firm or a team as well as suggesting potential consequences of human interactions. Thus it is possible to link isolated information of individual planners and make it accessible to the whole group. In this way the use of groupware plays an essential role as *coordinator of planning processes* in order to increase flexibility and the rate of change and improve collaboration between those carrying out work. Placing the groupware system into the existing computer environment was not simply done as a new component but it is used as a platform for integrating separate applications which were up to now only individual solutions. Thus closer linking between different modelling tools is reached and their interaction is supported by co-operation, integrated metamodel and by means of the implemented metamodel-coordinators.

The co-operation-coordinator assumes the overall system control of distributed modelling processes and links the planning tools to a monolithic toolbox which enables a continuous modelling of all design features from the shop floor. For instance effects from changed working organisation to the conception of material flow and information and communication structures are noted and processed. The result of groupware-based modelling is a consistent, comprehensive model of shop floor based on linking the different views of the manufacturing system. That leads to a new quality of planning a manufacturing system in contrast to conventional project realisations only for partial systems.

Contradictions and inconsistencies in parts of the manufacturing model can be recognised and eliminated as a consequence of integrating different models of particular tools based on an object-oriented metamodel. The groupware-based planning tool is capable of *giving feedback* on validation, revision, change or extension in order to recover lacks or faults effectively. It also takes consequences of improved or changed objects on other models in the system into account.

5.2 Implemented functions for improvement the collaboration

The computer-based execution of planning and modelling was realised in order to support and improve the collaboration between planners with regard to the design of a manufacturing system. For that following features were implemented in the toobox for distributed, co-operative modelling up to now:

- defining the project group and their chararcteristics
- asynchronous access on modelling data
- flexible consistenc verification, that means dependent from the planning situation, of the partial models concerning the whole model of manufacturing
- dynamic identification and making visible of the demand on coordination (scanning and monitoring)
- automatic initiation of communication processes through event control
- targeted, external communication about planning information.

Basis for the continuous execution of problem-related modelling functions is the object-oriented metamodel of the manufacturing which contains the technical objects and interrelations as well as the planning knowlege and the project data. Main demand on the conception of the metamodel was the integration of different views on the manufacturing system in order to reproduce the relevant characteristics. Information-technical basis of the different and required groupware functions is the groupware system Lotus Notes. In this connection, it is used (zum einen) as a modelling environment in order to create project teams, to define typical group characteristics or to flexible coordinate the planning work. (Zum anderen) it is used as an integration and development platform e.g. for integrating and implementing groupware functions which are still not complete. It is advantageous that the user interface is equal on all platforms. This is the reason if there was someting changed or extended, further expenditure to training courses is avoided since e.g. a new-designed mask behaves such as all others. A local network under Windows NT is the basis for communication and data exchange.

5.3 Procedure of the distributed, co-operative planning

Design and improvement of the shop floor in the field of manufacturing and assembly is carried out during a planning project where the procedure runs according to a rough plan concerned methodological reasons. Planning phases are partially graded in sequential planning steps which depend on one another logically. Moreover, parallel working phases are suitably supported in the framework of modelling and planning the manufacturing system. The modelling process is characterised by fuzzy ideas about possible solutions of the planning problem. Modelling itself depends to a large extent on creative working of the planning engineers. Therefore the coordination of separate modelling tasks is mainly based on a large number of communication processes among the team members.

- *Strategic planning*

The identification of a demand for changes is the starting point for every planning. Based on an assessment of the situation, it is recognised that a certain problem in manufacturing could be solved best through re- or new design of the company's elements. With this in mind, the underlying intentions and suppositions are summarised in a target concept. After that, a conceptual formulation to solve the relevant problem definition, which is the basis for planning, is developed as a result of preliminary analyses.

- *Creating a planning team*

Based on the target concept and the conceptual formulation of the problem field of re- or new design, a project team is created consisting of co-workers from different fields who are involved in planning. For that, information about the project group, the members participating in planning as well as the modelling tools used by the planners and thus implicitly the subtasks to be carried out are recorded and stored in the toolbox for distributed, co-operative modelling.

- *Planning study*

The main area of planning the manufacturing system is the ascertainment of an improved technical, organisational, personal and information-technical conception. Improvement of the manufacturing system starts with a business analysis the results of which are input data for the modelling tools. It is then completed by an intermediate report.

The later improvement of the planning models is divided into the following stages:
- system and structure planning
- rough planning (global planning)
- detailed planning (area planning)
- cost planning.

The real modelling tasks concerning those planning phases belong to the less-structured tasks which are coordinated dependent on the planning situation and use a large number of communication processes. The planning experts can work on the design of the manufacturing system asynchronously, simultaneously or in collaboration.

When a planner carries out his subrange planning asynchronously regarding the other planners, the partial models are transmitted to the object-oriented model after closing the modelling tool and then searched through for changes. All planners concerned by modifications in the planning model are informed and receive electronically the data changed according to their user view. After the completion of the planning phases, the model components processed simultanously are transformed to the object-oriented model. Incompatibilities and conflicts with interdependent model components of other model parts which are designed parallel can occur. The relevant project members are notified of this. In the following phase, they communicate about steps required to eliminate the inconsistencies and thus transmit their partial plans to consistent models. In addition, common work on a planning model is possible in which all planners involved have the same view of the model and synchronously carry out changes. Here only those planners must be informed who do not participate in the common modelling, but the partial models of whom are effected by the changes.

- *Project report*

The results of planning and modelling are summarised in a project report. It serves as documentation for decision-making regarding the execution of changes which are suggested, but not analysed or supported.

6 Advantages and dangers

The introduction of this *new technology* faces technical difficulties and problems in terms of contacts but these are not impossible to overcome. Some problem areas concern for example protection data security, monitoring and control by collaborators through documentation and interpretation of data and information, decrease in the charisma of eloquent team members, loss of power through sharing information as well as information overload and the phobia of technology by co-workers.

These difficulties result in a large number of *benefits and advantages*:
- improved quality management
- reduced meetings and duration of reviews
- significantly decreased expenditure with regard to paperwork
- reduced time spent on modelling
- better utilisation of expensive and specialised manpower
- improved collaboration between planners
- accelerated and simplified communication processes
- increased quality of planning
- dislocation of tasks among different planning departments without difficulties

7 Conclusion

The computer-based execution of distributed modelling tasks is now given by the co-operation concept developed here. Especially the dynamic aspect of collaboration is realised through this communication-supported workgroup concept. In order to support the group process of distributed, co-operative modelling by means of computers, the conceptions and technological possibilities of the research field Computer Supported Co-operative Work (CSCW) were used and then implemented in a toolbox.

The target of the group process concerning distributed, co-operative modelling is set as the improvement of collaboration between planners on the condition that the consistency of separate models based on the object-oriented overall model of manufacturing is kept. The consitency of different models is achieved by coordination processes which are realised on the basis of communication processes. Definite procedures are automatically triggered off dependent on activities of the project members concerning modifying, erasing or newly creating the common data stock. Automatic initiation of actions aims at the coordination of the interdependencies between separate tasks and at the same time at the maintenance the common data stock consistently. For that, messages are sent between the team members which are reacted to either by sending further messages or processing them.

Continuous controlling of the consistency of the modelling data leads to a reduction of planning time and an increase in the efficiency and additionally to an essential improvement of planning quality. Particularly the avoiding of misplanning caused by insufficient coordination between different planning experts is an essential criterion for using the toolbox developed here at single-item and small batch producers.

The continuous modelling work of experts from different manufacturing areas is mainly supported by
- identification of demand for co-operation
- initiating required coordination and communication processes.

Besides this modelling and planning work is improved by project-accompanying actions and functions such as group-scheduler, information sharing and various management functions. These newly developed units together with planning-accompanying components were implemented in a prototype of the overall tool modelling environment.

References

Dier, M., Lautenbacher, S., 1994: Groupware: Technologien für lernende Organisationen: Rahmen, Konzepte, Fallstudien. Computer Woche GmbH, München.

Dohmen, W., 1994: Kooperative Systeme: Techniken und Chancen. Hanser Verlag, München, Wien.

Hartmann, A., Herrmann, Th., Rohde, M., Wulf, V. (Hrsg.), 1994: Menschengerechte Groupware: Software-ergonomische Gestaltung und partizipative Umsetzung. Teubner Verlag, Stuttgart.

Hilgenfeldt, J., 1994: Werkzeuge für die computerunterstützte Organisationsgestaltung. Office Management 42 (1994)176, S. 42-45.

König, W., 1995: Wirtschaftsinformatik '95. Physica-Verlag, Heidelberg.

Luczak, H., Hinz, S., Quaas, W., 1995: Experte Mitarbeiter: Strategien und Methoden einer mitarbeiterorientierten Gestaltung und Einführung rechnerintegrierter Produktion. Verlag TÜV Rheinland, Köln.

Mertens, P., 1995: Integrierte Informationsverarbeitung 1. Gabler Verlag, Wiesbaden.

Wedekind, H., 1994: Verteilte Systeme. BI-Wissenschaftsverlag, Mannheim u.a..

The work described is funded by the German Research Association (DFG) - Scho 540/1-2.

Part 3   Strategic Business Planning

A Proposal Approach for Strategic Probe: A Scanning Information Support
K. Rouibah, H. Lesca

Integrating the Strategic and Technical Approach to Business Process Engineering
J. L. G. Dietz, H. B. F. Mulder

A Proposal Approach for Strategic Probe:
A Scanning Information Support

Kamel Rouibah[1] and Humbert Lesca[2]

[1]Centre de Recherche Appliquée à la Gestion   (E.S.A), Institut pour la Production Industrielle (I.P.I), Ecole Nationale Supérieure de Génie Industriel (E.N.S.G.I), 46 Avenue Félix Viallet , 38031 Grenoble , France, Kamel.Rouibah@ensgi.inpg.fr

[2]Centre de Recherche Appliquée à la Gestion   (E.S.A), CERAG, Department of Information System. (ajouter l'adresse de l'ESA)

Abstract. Environmental scanning and information interpretation provide crucial information on which strategic decision making is based. This paper discuss a new concept termed « Strategic Probe » that allows organisation to be secure from environmental surprise. It deals with a kind of information called «weak signals ». The quality, and therefore the success of strategic probe depends on the relevance, validity, accuracy, and timeliness of the data gathered and their interpretation. So this paper introduces a model which tend to facilitate building a coherence and meaningful image (knowledge) from widespread information. Details about interpretation model are presented.

Keywords. Strategic Probe, Weak Signals, Strategic business planning, Business Intelligence, Business Environmental Scanning System, Strategic Management, Information Structuring, Information link, Creativity, Emerging sense, Reasoning

1 INTRODUCTION

If designed properly, conventional MIS can be quite successful in helping managers to identify organisation's internal strengths and weaknesses. However, these systems are not able to provide support to manager's need to be aware of *external changes* which lead to threats or opportunities. Such support does not exist yet. Our paper intends to give insight information about the problem.

Effective strategic response to environmental changes requires a clear perception of events and trends in organisation's external environment. In France, managers and organisations form these perceptions through a process under the name of «la veille stratégique » called in this article as «strategic probe ».

Our attention in this paper is devoted to the potential computerisation of strategic probe and interpretation activities (using tool), as they relate to the external environment. The reasons that make us focus on external environment are:
1. Then higher the level of management is, then more attention is given to the external environment.
2. Interpretation of the external environment is usually more complex than the internal environment.
3. Nowadays most existing decision support systems do not attempt to support manager's effort in scanning and interpreting external environmental changes. It is hoped that this paper will provide some guidelines for such a support.

This paper is organised as follow: first it discusses the process of strategic probe. Second, it introduces the problem of emerging sense from gathered information. Third, it presents the resolution approach. Finally, conclusions are provided.

2 MODEL OF STRATEGIC PROBE

In order to design any successful computer based information system to help manager gain clear perception of the external environments, it is first necessary to understand the strategic probe process.

Definition

The « Strategic Probe », according to several researchers (Ansoff 1975), is a systematic approach to a major, and increasing in importance responsibility of general management. It tends to position and relate the organisation to its environment in a way that will assure the organisation's continued success and make it secure from surprises.

Strategic Probe is a component of business information system. It is known in the literature as environmental scanning (Fahey, King, and Narayanan 1981), Strategic Information Scanning System (Aaker 1983) or even business environmental scanning (Narchal, Kittappa and Bhattacharya 1987). Our approach differs from the ones mentioned above because it deals only with a kind of information, called « weak signals » (Ansoff 1975) which are picked up from external environmental organisation (see Figure 1). These information concern external events and actors, and their aims are turned to the future instead of the past.

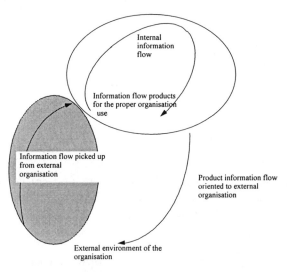

Fig. 1: Information flow

Objective of Strategic Probe

The process of strategic probe has recently started to be studied in France, whereas it is more developed abroad (Jain 1984). It is the process by which a company keeps being informed in an anticipatory way about opportunities and threats which occur in its socio-economic environment, by collecting anticipate information (Lesca 1986, Lesca 1994), instead of calculating statistical forecast (Fahey, King and Narayanan 1981).

Strategic probe plays a role for business as same as a radar for a ship. It aims to inform managers early enough, about interesting opportunities and threats so that they become able to cope with them.

Phases of Strategic probe

The process of strategic probe can be divided into five phases (see Figure 2). The first phase can be executed either informally by individual executives or formally carried out by an organisational unit. The second phase consists of routing weak signals which are collected from the outside (external) to the inside (internal) of the organisation. Most of the time, you do not know to whom and how to route the collected information which thus have every chance of being lost. The third phase transforms weak signals from a chaotic situation to a structured one, in order to be useful for executive's action. At the end of the previous phase, information can be significant. In this case, they conduct to take actions (phase 4). Otherwise, information may be inaccurate or imprecise: so they need to be completed. Feedback loop is involved (phase 5) to return to the first phase, but an information research is developed in a specified way.

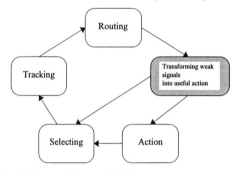

Fig. 2: Phase of Strategic Probe

Each phase raises many questions, but the third one is the most crucial and difficult one. So we will focus on it. Other phases can also be computerised, but this topic is outside the boundaries of this paper.

Different Kind of Strategic Probe

Strategic Probe includes different kinds of scanning such as competitive, commercial, and technological. Specialised news papers tend to focus more and more on the technological, while our paper attempts to investigate the competitive scanning.

3 WORK CONTEXT

We firstly specify the work context in order to facilitate the approach comprehensiveness developed in the next section.

Established Fact: Difficulties to Treat Weak Signals

Our research team has conducted many researches. They show slow development of the strategic probe. Here are the reasons:

1- Nature of the Information Involved.

The characteristics of each information make their management very difficult. Information is (El Sawy 1985, Lesca 1986):

Anticipatory. It must inform management about the changes occurring in the external environment which are not realised yet.

Qualitative. In most cases, it do not consist of numbers which record or extrapolate the past. Therefore, information may be picked up or overheard, in many shapes such as written, verbal, image, in meetings, exhibitions, conferences, etc.

Indistinct. We are not almost certain of picked information. It could be swamped in the « noise » or even be deliberately contaminated or distorted by an other party.

Fragmentary. Each information, taken a part, is insignificant and suspect. However, it acquires significance gradually when it is gathered with other information.

2- Lack of Experience

Our researches reveal the very poor training given to current and future managers on the use of information towards strategic ends. In other words, organisational executives ignore environmental changes.

3- Lack of Information Tools

Tools and methods which treat this kind of information are inappropriate. DSS like RDSS (Toda, Shintani and Katayama 1991), CLSS (Yadav and Khazanchi 1992) Leximappe (Courtial 1994) are not appropriate to deal with weak signals. In addition methods like Quick (Nanus 1982), Leap (Preble 1982) Delphi method brainstorming, etc., require a group session and also experts coming from outside of the organisation. So they did not allow to create effective and easy knowledge in short time.

Deduction

We presume that manager's behaviour towards environmental scanning would be very effective if decision support (DS) is developed. We use the term of decision support (DS) to assign a method, a methodology or a practical tool.

Problem

The problem can be expressed as follow : let's suppose that fragmentary scattered and suspected information are collected. Each information has no great significance when taken into account in a lonely way.

How can weak signals be classified in a manner to help users to seize new opportunities or to avoid threats ?

How can a chaotic situation be moved to a structured one ?

How can useful meaningful signals be created from a jumble of widespread weak signals for senior management action?

What simple and efficient tool can help companies managers to find their way through a chaos of weak signals ?

Because there is neither methodology nor tool which allow us to develop sense from a jumble of widespread information, we consider the problem, according to (Toda, Shintani and Katayama 1991), as « an ill-structured problem » for which a solution procedure is not completely specified.

Face to this reality and to answer the questions above, our approach intends to structure the ill-structured problem.

Purpose

The purpose is to establish an approach and a tool in order to make managers involved in an environmental scanning through methods other than speeches. Moreover it enhances manager's comprehensiveness so that occurring changes are anticipated.

4 MODEL OF SOLVING APPROACH

According to Simon's decision making process (Simon 1977) our problem deals with « intelligence phase » which consists on problem identification and formulation. Therefore, when the DS supports problem structuring, it is more important that it supports structuring information. The information structuring consists of integrating and transforming pieces of insignificant collected information into the appropriate one for useful action.

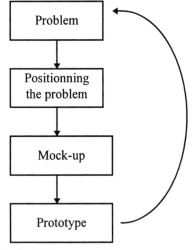

Fig. 3: Solving approach

This process of integration and transformation can be only performed by trials and errors due to the ill-structuredness of the problem. So we identify the problem, it can therefore be formulated in several ways, but this must be consistent with the process chosen to represent the problem, not with an objective that may not have been formulated or may not exist yet. We then elaborate a mock-up « a surface » approach which will be implemented as a prototype. Later on, this prototype can evolve or even completely change.

As pointed above, creating an approach and a tool that could facilitate meaningful signals creation is not so easy. So, we turned towards publications which, on one hand, are concerned with heuristics (Sugiyama and Toda 1985, Smith 1988), and in the other hand, with creativity (Couger, Higgins and McIntyre 1993).

Creative Methods Contribution in Our Model Conception
- *Categorising information.* In order to facilitate information use, researchers propose to categorise information (McKenney and Keen 1974, Couger, Higgins and McIntyre 1993).
- *Making Association.* According to Mednick in (Hoggarth 1980), to be more creative, it is useful to bring information together so that images could emerge from chaos.
- *Elaborating multiple arrangements.* According to Kanter in (Couger, Higgins and McIntyre 1993) « kaleidoscopic thinking » allows us to take existing information: « twist them, shake them, look at them upside down or from an other angle or from a new direction ». This generates entire new patterns and consequent set of actions. K. J. method (Kawakita 1975) uses the same mechanism called blind arrangement in order to produce a clear image from a jumble of widespread information or items.
- *Visualising association.* Individuals are more attracted by visionary images. According to (Moles 1990), thinking is visualising.

The Approach

Our approach, based on (Valette 1993) work, consists on elaborating a graph called « Puzzle » that presents similarities and differences with cognitive map (Calori, Johnson and Sarnin 1994). Our puzzle approach is analogical with the jigsaw puzzle because of the reasons above explained. According to several authors (Toda, Shintani and Katayama 1991), our approach is a combination of two steps consisting in categorisation and graph elaboration:

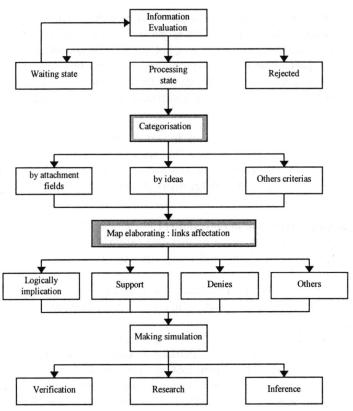

Fig. 4: Mock-up of approach

1- Categorising weak signals.

Classifying collected information is useful to facilitate their use (Behling, Gifford and Tolliver 1980). It can be done:

By actors

It depends on the considered actors such as: suppliers, customers, competitors, distributors, regulators (government), lobbies, etc.

By current status

This could be « rejected », « waiting state » or even «processing state». The criteria used to assess weak signals are:
- Validity. The degree to which the information reflects the reality according to you (information objective / subjective),
- Familiarity. The degree to which the information, at the time of its receipt, is or not already known by you (known / unknown),

- Timeliness. The availability of the information at a time that this information could be used (information was in time / was not),
- Actionable. The ability and the plausibility to take action on the basis of this information (we can do lot of things / we can not do anything),
- Completeness. The comprehensiveness of the information, i.e. the degree to which all the essential elements have been dealt with, according to you (full (sufficiently complete, / fragmentary),
- Surprise. The degree to witch the information is unanticipated, and runs counter what you thought was true (low surprise (in online with my opinion) / high surprise (is in contradiction with my opinion),
- Accuracy. The preciseness of the information. (very precise / not very precise)

We computerise a weight for each information retain, that is a weak signal. It consists of a combination of the above criteria. It allows us to make simulation during next phase.

By attachment fields

Those are R&D, production, application and product. Each of these concepts is related to :
- sellers entries with a new technology (originally out of the industry),
- sellers entries with a similar technology (originally out of the industry),
- buyers entries: new customers groups for the same applications,
- buyers entries: new customers group for the new applications,
- sellers exit: transfer to another competitor,
- sellers exit: liquidation,
- buyers exit,
- sellers investments: product R&D,
- sellers investments: process R&D,
- pricing strategies,
- international development strategies,
- delocalisation strategies,
- horizontal alliance between sellers,
- horizontal alliance between buyers,
- horizontal alliance between suppliers,
- vertical alliance sellers-buyers,
- vertical alliance sellers-suppliers,
- acquisitions and mergers,
- sellers diversification (focus on the industry or multi-industry),
- government interventions on prices, regulations, on restructuring, quotas, forms, laws,
- influences on various lobbies.

So every observed weak signal has to be attached to one or more of the above criteria.

2- Elaborating significant representation

These representations, which are meaningful maps, are named « Puzzle ». A puzzle is a graph where:
- nodes are small sentences corresponding to gathered weak signals,
- and edges link different nodes which may be Logically Implication, Support, Denies, Presuppose, (Lee and Lai 1991, Moles 1990, Toda, Shintani and Katayama 1991).

So atomised information becomes an intelligible representation. Puzzle elaborating is a real act of creativity. After the information selection phase, the end user develops a mental effort so that different information are connected as presented in the following picture (see figure 5) :

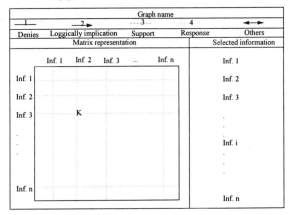

Fig. 5: Graph construction

Element (i, j) = k, from the upper matrix with k = 1, 2, 3 ... n, is the possible links between the information.

After reading all the information related to the interest topic, the end user identifies possible existing links between information. Then he fills the matrix above. Going from a matrix representation to a graphical one, allows us to (see Figure 6) (Lesca 1992):

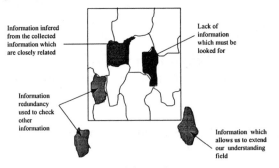

Fig. 6: Example of Puzzle elaborated

Provide significant representations. The arrangement of the information, or pieces of puzzle must be easily movable. This requires an appropriate computer tool which enables the creation of links between fragmentary information. When it is done, it allows us to go from atomised and insignificant information to intelligible representations.

Make simulations. An elaborated map may not be conformed to any preconceived model. Thus, it is necessary to create several representations in order to find the most adequate one to perceive the most adequate and sensitive managers reality. Those simulations are made according to weights affected to information. Note that only information with weight Pi $\geq \alpha$ will be visualised; α is the weight fixed by the end user.

Provide verification of reassembled information coherence. By linking information, cross-checking and validation of the stored information is progressively completed.

Infer inaccessible information. It is possible to make inferences from information stored in the data base. These information can not be accessible if they are not brought together. Therefore, one should take this as a working hypothesis which need validation.

5 CONCLUSION

We have presented an approach (a mock-up) for emerging sense. The first step (storage, information structuring and retrieval) is now implemented on Lotus Notes which is an environment for work team, developed by Lotus. It is also on Microsoft Relational Data Base Applications ACCESS, developed by a DESS student. Future researches will be oriented in the approach improvement especially about the second step:
- how to facilitate inaccessible information inference?
- how to orient actions?
- how to support making hypothesis?
- how to make arguments?
- how to minimise crossing links between information of a graphical puzzle?
- how to combine Lotus Notes and COPE capabilities in order to best develop our software? Cope is decision support system developed by (Ackermann, Cropper and Eden 1992), used for elaborating cognitive map.

186

References

Aaker, D. A. (1983) Organising a strategic information system. *California Management Review*, vol. 25, n' 2, pp. 76-83.

Ackermann, F. Cropper S. and Eden C. (1992) COPE user guide .

Ansoff, H. I. (1975) Managing strategic surprise by response to weak signals. *California Management Review*, winter, 1975, Vol. XXVIII, n' 2, pp. 21-33.

Behling, O., Gifford W. and Tolliver J. M. (1980) Effects of grouping information on decision making under risk. *Decision Science* n' 11, pp. 272-283.

Calori, R. Johnson G. and Sarnin, P. (1994) CEOs' cognitives maps and the scope of the organisation. *Strategic Management journal*, vol. 15, pp. 437-457.

Couger, J.D., Higgins L. F., and McIntyre S. C. (1993) (Un)structured creativity in information systems organisations. *MIS Quaterly*, December 1993, pp. 375-397.

Courtial, J. P. (1994) A coword analysis of scientometrics. *Scientometrics* n'.31, pp. 251-260.

El Sawy, Omar (1985) Personal information systems for strategic scanning in turbulent environments: can the CEO go on-line?. *MIS Quaterly*, March 1985, pp. 53-60.

Fahey, L. King, R. W. and Narayanan V. K. (1981) Environmental scanning and forecasting in strategic planning: The state of the art. *Long Range Planning*, vol. 14, February (à vérifier), pp. 32-39.

Hoggarth, R. (1980) *Judgement and choice* .Ed. John Wiley and Sons (250 p)

Jain, S. C. (1984) Environmental scanning in US corporations. *Long Range Planning*, vol. 17, n'2, pp. 117-128.

Kawakita, Jico (1975) The K. J. Method. *Kawakita Research Institute*, Tokyo, 12 p.

Lee, J. and Lai, K. Y.(1991) What 's in Design Rationale? *Human Computer Interaction*, Vol. 6, pp. 251-280

Lesca, H. (1986) *Système d'information pour le management stratégique de l'entreprise*. Paris, McGraw Hill, 146 p.

Lesca, H. (1992) Le problème crucial de la veille stratégique: la construction du puzzle. *Revue Française de Gestion*, April, pp. 67-71.

Lesca, H. (1994) Veille strategique pour le management strategique état de la question et axes de recherche. In *Economies et societés*, série science de gestion, n°20, 5/1994, p.31-50.

McKenney, L. J. and Keen P. G. W. (1974) How managers' minds work. *Harvard Business Review*, vol. 3, n' 52, pp. 79-90.

Moles, A. (1990) Les sciences de l'imprécis.

Nanus, B. (1982) Quest: quick environmental scanning technique. *Long Range Planning*, vol. 15, n'2, pp. 39-45.

Narchal, R. M., Kittappa, K. and Bhattacharya, P. (1987) An environmental scanning system for business planning. *Long Range Planning*, vol. 20, n' 6, pp. 96-105.

Preble, J. F. (1982) Future forecasting with LEAP. *Long Range Planning*, Vol. 15, n' 4, pp. 64-69.

Simon, H. A. (1977) The new science of management decision. Prentice Hall, Englewood Cliffs.

Sugiyama, K. and Toda, M. (1985) Structuring information for understanding complex systems: A basis for decision making. Fujutsu Science Technological Journal, n' 21, vol. 2, June, pp. 144- 164.

Toda, M. Shintani, T. and Katayama, Y. (1991) Information structuring and its implementation on a recherché decision support system. *Decision Support Systems* n' 7, pp. 169-184.

Smith, G.E. (1988) Towards a heuristic theory of problem structuring. Management Science, vol. 34, n' 12, pp. 1489-1506.

Valette, F. (1993) Le concepte de Puzzle: coeur du processus d'écoute prospective de l'environnement de l'entreprise. Thèse, Ecole Superieur des Affaires.

Yadav, S. B. and Khazanchi, D. (1992) Subjective understanding in strategic decision making: an information systems perspective. *Decision Support Systems* n' 8, p. 55-71.

Integrating the Strategic and Technical Approach to Business Process Engineering

Jan L.G. Dietz and Hans B.F. Mulder

Department of Information Systems, Delft University of Technology, P.O. Box 356, NL-2600 AJ Delft, The Netherlands, secr@is.twi.tudelft.nl

Abstract. Executives are experts in managing their business systems, much in the same way Formula One drivers are experts in driving their cars. However knowing the behaviour of cars doesn't tell one how they are constructed and could be improved with new technology. The same holds true for Reengineering Business Systems through Information Technology. If management wishes to redefine the organisation, technical skills and knowledge of reengineering information systems are required to realise these objectives. Therefore Reengineering the Organisation will have to integrate a technical and a strategic management approach. In this article the Theory of DEMO is presented as a fundamental new perspective for Reengineering Organisations through IT.

Keywords. Dynamic Essential Modelling of Organisations, Reengineering, Strategic Management

1 Introduction

Market turbulence and technological developments are changing organisations and the way in which business is conducted. Business Redesign is necessary in order to act constructively. In this Information Technology is seen as an enabler to achieve new forms of organisation (Hammer, 1990; Scott Morton, 1991). After losing the interest of management, employees and consequently consultants, technical methodologies such as Information and Business Systems Planning are being replaced by IT-induced strategies. These current approaches to Business Redesign largely draw from economic theories, such as industrial organisation economics for strategic management (Porter, 1980). Models that derive from Porter's Value Chain (Porter, 1985a) have appealed to the interest of management in general and of scientists, especially in the field of Business Redesign (Scott Morton, 1991). Models as these have made a huge contribution as a starting point for Business Redesign, but they lack the ability to support consistent management decision making and communication in the further and more specific aspects of systems design, for one simple reason. Strategic Management is about allocation and growth of organisational resources and not about a powerful change approach that can bring radical improvements by redesign of business, information and data systems.

Therefore a new perspective on organisations as well as on the practice of IT has to be established, which provides an approach and tools that address the envisioning, enablement, or implementation of radical change (Davenport, 1993). However, the popular management literature has created more myth than practical methodology (Davenport, 1994; Dietz, 1994). In this article a revolutionary foundation for Business Redesign through IT and a more coherent approach is presented than currently applied in BPR-methods. This way of thinking is based on the informational theory of Dynamic Essential Modelling of Organisations (DEMO) by Dietz, which captures the essence of organisations by analysis of transactions from a Language/Action perspective. On the principle that complex processes can be decomposed into discrete transactions, the 'primitive' building blocks are provided for translating Business Redesign into information systems specifications, thus providing a grounded approach for Business Redesign through Information Technology (Keen, 1991).

2 The Problem of Strategic Management Foundations for Business Reengineering

The history of information systems design is a history of borrowing from other disciplines. This borrowing began when people suggested similarities between business processes and computer software processes. From this perspective, business processes were viewed mechanistically as if they were computer programs. The first systems design methods were an extension of software engineering principles. Other concepts were borrowed from organisational science to address important social factors in implementing IT-projects. More recently, ideas from microeconomics (Porter, 1980; Porter and Millar 1985) have been borrowed in the development of IT-induced strategies and the Business Redesign model (Scott Morton, 1991). Although the use of methods, such as Yourdon's System Development and Porter's Value Chain have advanced and contributed to parts of business systems design, the underlying assumptions, reasoning and common paradigm of these sources lead to difficulties in the field of realising Strategic Management Reengineering Objectives.

System analysts or designers, or whatever title is used to describe the role of intermediator between management and software engineers, will have to understand both the strengths and limitations of these non-informatical theories in their profession. Examples of purely informatical methods, are DEMO and the Flores-Winograd workflow model. These methods focus on communication processes to understand the business, rather than starting to examine the current organisational structures or documental flows. We will not elaborate on all sorts of problems that can arise when designing systems on non-informatical foundations, but we will give an impression of the pit-falls when applying technical or strategic management approaches.

2.1 The Technical Approach

Originating from the 1950's the common paradigm of software engineering is seeing information and data systems as lines of code or as tables and relations, nowadays. Software engineering principles show a bottom up approach with regard to data collection, Modelling and design.

When an analyst takes as 'given' the validity of current business processes, the starting point of IS design and development is data collection rather than business processes. Data collection methods are mostly based on interview techniques, work observation or analysing current documental flows. Then a business process model is composed of the gathered and selected 'relevant' data followed by sophisticated Modelling principles. Lastly systems are defined, which can be done by rigidly executing techniques such as Create/Use matrices.

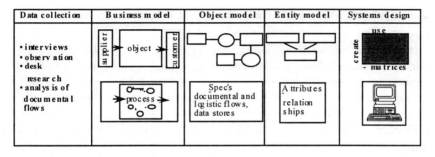

Data collection	Business model	Object model	Entity model	Systems design
• interviews • observation • desk research • analysis of documental flows	supplier → object → customer process	Spec's documental and logistic flows, data stores	Attributes - relationships	use - create - matrices

Fig. 1 Overview of Systems Development

The ultimate goal of this paradigm seems to be the development of information systems. Applying the techniques lead to programmable representations of the organisation. This implementation driven approach focuses on the documental and informational level for increasing the efficiency of the current organisation.

The technical paradigm was valuable within the given time frame of early automation. Programming was complex, time consuming and required expensive and highly trained professionals, which implied that technical implementation had its limitations. It was key not to limit the organisation to this immature state of technology. So one of the rules of thumb became: automation has to fit precisely in existing organisational procedures and documental flows. This rule is contrary to the approach of business redesign.

The software engineering approach was a major step forward in systems development, but borrowing from software engineering without regard to or being aware of its goals and assumptions has it risks for systems design. Methods for Business Systems Planning based on these premises are considered too technical and complex to be fully understood and executed by management and employees. Because of the need for consultancy and the time and effort of the organisation members to specify all processes and data flows on paper, information planning comes at a too high cost.

Although formal methods provide in-depth techniques and consistency checks for information analysis, formal methods seem to have gone out of favour with management, employees and consequently consultants, but the arguments for systems planning and design remain. For this reason some organisations have introduced light versions of information planning methods. These so called light versions give directions regarding the contents of an information plan, but do not suggest the use of method. The light versions allow people freedom in ways to think and work, but they do not provide a foundation (i.e. common reference) for analysis, design and implementation.

2.2 The Strategic Management Approach

The rules of competition and of organisation changed in the 1980's by two major trends: Market turbulence and IT developments [Scott Morton, 1991]. Many of the constraints imposed by these market forces cannot be anticipated, and thus enormous uncertainties and ambiguities arise concerning the appropriate responses to these forces. These uncertainties, in turn, had seemed to suggest that a firm's performance depends on its good fortune more than on its quality of management. These ideas were contradicted by several research traditions in microeconomics, notably Industrial Organisation (IO) economics, which argues that organisations both shape and respond to market forces, and that a firm's performance ultimately depends on this mix of action and response. This IO conception of competition has shown to be relevant to business redesign through information technology. The IO version has been incorporated into popular theories of strategy, by Porter among others.

In IO economics industries are the unit of analysis. Porter (1981) states that this basic concept of IO, first developed in 1939 by Mason and Bain, offered strategic management a systematic model for assessing competition within an industry. In this model, returns to firms are determined by the structure of the industry. It was Porter's contribution to lift the deterministic limitations of the Mason/Bain paradigm by showing Competitive Strategies that change the structure of a firm's industry and thus gaining superior performance (Porter, 1980). Thus Industrial Organisation economics offers strategic management a systematic model for assessing and structurally changing competition within an industry to gain superior performance (Porter, 1981). Later the Value Chain was presented as a concept to achieve Competitive Advantage (Porter, 1985b).

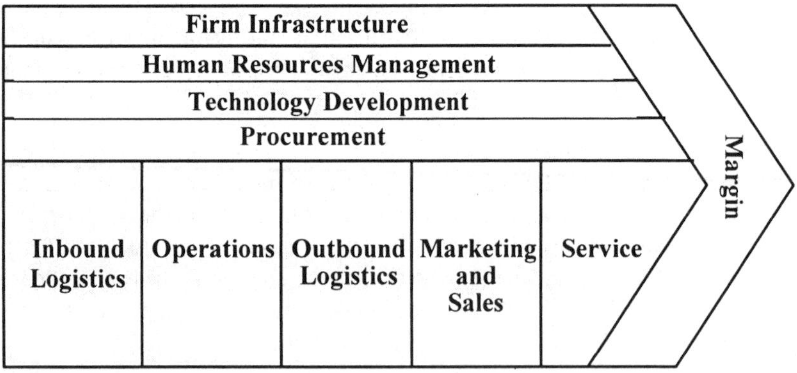

Fig. 2: Plot of the Value Chain

More recently, the ideas of IO economics (Porter and Millar, 1985a) have been borrowed in the development of IT-induced strategies for gaining competitive advantage, which is a starting point for Business Redesign. Business Redesign advocates a revolutionary change of structure and processes of the organisation in

order to act constructively. Information Technology is seen as an enabler to achieve a new organisation (Hammer, 1990; Scott Morton, 1991; Davenport, 1993). Five primary concepts can be identified that make up Redesign :
1. A clean slate approach to organisational design and change.
2. An orientation to cross-functional business processes, or how work is done.
3. The need for, and possibility of radical change in performance.
4. Information technology as an enabler of change in how work is done.
5. Changes in organisational and human arrangements that accompany change in technology.

It is the combination of concepts that is new in Business Redesign (Davenport, 1994), not the concepts considered individually. Venkatraman defines this IT-enabled business reconfiguration as consisting of five stages, in which the first two stages, automation of isolated activities and building an infrastructure, are the foundation for the other three non-sequential revolutionary stages (Scott Morton, 1991). Stage three, business process redesign, concerns rethinking and integrating the most effective way to conduct business internally. Stage four, business network redesign, includes suppliers, customers, or anyone else who can contribute to the firm's effectiveness. In stages three and four the organisation remains within the same industry, which has a resemblance to industrial organisation economics. The fifth stage is business scope redefinition, which sets out to exploit new technology in the marketplace and products, much like a Schumpeterian revolution.

In contrast to the technical approach based on software engineering concepts, strategic management is characterised by a more top-down approach. Venkatraman puts it as follows (Scott Morton, 1991):

> A fundamental tool, as IT is best deployed by those with enough vision to see what it can realistically mean to the organisation. In short, general management, and preferably senior general management, should be in direct control. This predicated on a fundamental logic that IT is to be increasingly viewed from a general management (or strategic management) perspective in addition to the traditional perspective of the information systems (IS) function. This reflects a logic that effective exploitation of IT, could involve significant change in organisational strategy, management structure, systems, and processes that can not be accomplished from a functional perspective.

The differences between the strategic management and the traditional technical perspective are evident: The guiding principle in strategic management is on alignment of the Business and IT strategy and not IT for implementation.

2.3 Fundamental Difficulties with the Strategic Management Approach

The emphasis of strategic management is, of course, on strategic issues, but how are practical questions solved like 'How to distinguish revolutionary opportunities for redesign ?' and 'How to model the way of rethinking the internal and network processes ?' and 'Which techniques are suitable ?". Davenport (1993) states about analysis precluding to the design and implementation the following:

> Inasmuch as understanding current processes is an essential element of all traditional process-oriented approaches, we look to these methods for tools and techniques that might be used to improve existing processes in an innovation initiative. Process diagrams, such as flowcharts, are among the most useful of these tools. Traditional approaches use flowcharting to understand and communicate about processes. The cost build-up chart, another commonly used tool, can be used to reveal process cycle time. In all cases, care must be taken to ensure that the tools and techniques selected fit the overall objectives of the innovation effort.

> None of these traditional improvement approaches is likely to yield radical business process innovation. Although they may share characteristics with process innovation, all begin with the existing processes and use techniques intended to yield incremental change; none addresses the envisioning, enablement, or implementation of radical change.

A management interpretation of the practical implications of the Business Redesign approach is given by (Scott Morton, 1991). When examining the Value Process Model of MacDonald, the following statements are interesting:

> Value Chain diagrams are frequently used to illustrate the transformations that have taken place in organisations. But many practical problems arise when attempting to develop value chains for an organisation and then using the value chains as the basis for strategic review and business process redesign. In practice, some types of industries, particularly government and service industries, do not easily relate to these headings.

Remarkably the references to theories or methods that support management are limited only to (Porter, 1985b). Porter's Value Chain provides general principles for improving products, customer and supplier relationships, and creating barriers for substitutes and competitors. The Value Chain is adequate to present an overall picture of the organisation and its environment, but this perspective is too general for realising Reengineering Objectives. In practice management supports projects which aim to bring organisational change based on the ideas that seem valid at that time.

2.4 Conclusion

One observes that Strategic Management approaches, like Porter's Value Chain however popular they may be, are lacking practical methodology and suitable techniques for Reengineering. Some difficulties of the strategic management approach for reengineering the firm are :

- Modelling of different types of industries

  According to the Value Chain the first model of an organisation needs to be decomposed to the levels of inbound logistics, operations, outbound logistics, sales and marketing, and service. Originally (Porter, 1985b) the value chain focused on production. Later (Porter, 1990) services, such as information processing, were positioned in activities, such as outbound logistics. As (Creemers, 1993) notes 'Services are seen as adding value to goods production. This is not a satisfying situation. How, for instance, can a bank or an insurance company profit from the value chain approach without using terms such as inbound logistics or without adding value to manufacturing processes ?'. Thus grounded reasons for starting with a functional decomposition of inbound logistics, operations, outbound logistics, sales and marketing, and service remain unclear.

- Alignment of Business and IT strategy

  To realise Strategic Management Reengineering Objectives, concepts as 'inbound logistics' have to be transformed into the details of information system design. Simple IT principles and notions such as electronic data interchange or bar-coding to improve 'inbound logistics' are all too general to be considered the advocated change in organisational strategy, management structure, systems, and processes, which can not be accomplished from a IS functional perspective. When attempting to realise Engineering Objectives on the basis of a functional (re)design of the Value Chain, one will experience that functional decomposition doesn't lead to specifications for technical redesign; knowing the behaviour of business systems doesn't tell one how it is constructed and could be improved with new technology.

- Decomposition

  Decomposition is poorly supported to one level of abstraction, such as Operations or Sales. Methods for how decomposition is achieved and consistency is checked seem to be missing in the Value Chain Theory. In practice this leads to using technical techniques for systems development, which are more suitable for implementation of IT than Reengineering the Organisation through IT.

Clearly the starting point of the Value Chain is strategic, but the management decision making and communication is not supported in the additional and more specific process of realising Strategic Reengineering objectives. The idea of having separate strategic management methods and separate techniques, based on software engineering principles, for realising the strategic management reengineering objectives has evolved to be self-evident. This division of methods for management and reengineering have rooted deeply. Therefore it is important to offer a new perspective which distinguishes and integrates the levels of business and information.

3 Dynamic Essential Modelling of Organisations

3.1 Redesign with DEMO

In this section a theory and analysis method are presented which provides the tools to examine business transactions and as well as a revolutionary method for realising business redesign through IT. Keen (1991) points at the conversion for action approach as a new approach to analyse business processes. According to Keen one of the distinctive strengths of this approach is that it provides a customer satisfaction base for targeting IT applications within and across the phases of the work flows. Many work flows require cross-functional activities. This can lead to breakdowns in coordination and affect customer satisfaction by creating delays, lost information, or misunderstandings. As an exponent of the language action perspective Keen refers to Flores and Winograd (1986), who built the BDT-method and subsequently the Action Workflow products on the principle that a basic business process is a four-part "action work flow" (proposal, acceptance, performance and satisfaction phase), so that complex processes can be decomposed into workflow loops. These loops provide the 'primitive' building blocks for translating Business Redesign into information systems specifications and visa versa. From a management perspective every activity inside the customer/performer loop - and every information or processing system - can damage or enhance customer satisfaction. A comparable approach is proposed in (Dewitz and Lee, 1989).

Reengineering is a powerful change approach that can bring radical improvements in business processes. However, the popular management literature has created more myth than practical methodology regarding reengineering (Davenport, 1994; Dietz, 1994). When examining current literature on Business Redesign, analysis methods and techniques that support Business Redesign are exceptions (Reijswoud and Rijst, 1995). One of them is Dynamic Essential Modelling of Organisations by Dietz et al. DEMO draws on the philosophical branches of semantics and scientific ontology. An other major source has been the mathematically defined rigid approach to the notions of discrete systems and Modelling system dynamics, as described in (Dietz, 1989; Dietz, 1991). The 'conversation-for-action' approach as implemented in the Action Workflow products (Medina-Mora et al., 1992) has been one of the major sources of inspiration for the development of DEMO. The Language/Action perspective argues that the core of an organisation is the network of discrete recurrent conversations. Applying the language/action perspective has generated numerous insights both in academic (Dietz, 1991) and applied setting (Medina-Mora and Denning, 1995).

The idea behind Business Redesign is defined as that more effective usage of IT can be achieved if the IT applications would be designed along with a redesign and reengineering of the business itself. To support the practical realisation of this idea, one needs a model of the organisation which shows the essential characteristics of the business and which abstracts from the way in which they are realised by means of information technology and organisational structures (Dietz, 1994). In much the same way as a documental model is of little use for studying informational issues, an informational model is not very well suited for studying the real business issues from the informatical point of view. These issues come into sight when we extend our focus to the essential level.

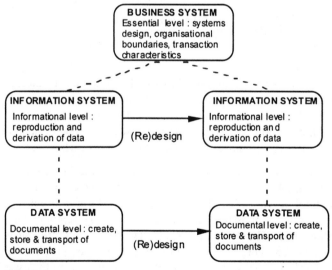

Fig.3 Redesign

Figure 3 shows that for any organisation there exists one documental model, one informational model and one essential model at any moment. In principle though, one may conceive of a number of documental models, all realising the

198

same informational model and one may conceive of a number of alternative informational models realising the same essential model. The transfer from the essential model to an informational model is what (Re)design is about. In principle there are a number of informational models, all realising the same essential model. Otherwise said, there is a freedom of choice, and consequently a choice problem (Dietz, 1994).

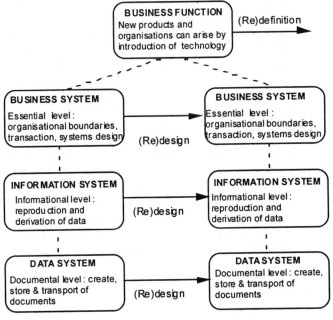

Fig. 4 Redefinition

At the documental level an organisation is viewed as a system of actors that produce, store, transport and destroy documents. At the informational level one abstracts from the substance and the syntactic aspect in order to focus on the semantic aspect of information. What one observes now is a system of actors that express and receive messages (semantic meanings) to and from each other.

This is the level where most current technical methods and techniques (e.g. the DFD and the ER-model) aim to be helpful. In DEMO these two levels are extended with a third level, called the essential level. As a consequence of abstracting from informational issues, all actions concerning reproduction and derivation are eliminated. The term "essential" has been used previously in information notably in (Yourdon, 1989), but does not differ from a logical model, which aggregates physical data processes. To explain the real value of the essential model, one should compare the actual way of carrying through a financial transaction of a bank some 40 years ago with the way they are carried through with information technology, nowadays. The essential model is the same in both cases, but all less abstract models (e.g. those expressed in DFD's) differ. Another characteristic of DEMO is discrete decomposition. Contrary to Modelling methods in the technical approach which often lead to an uncontrolled, complex, subjective and paper consuming explosion of objects and relations, which makes these models almost inaccessible, DEMO provides intuitive and effective models. A comparison between existing speech-act based methods, such as used in Action Workflow products, shows that DEMO offers a distinction in models to differentiate the essential from the informational and documental level. Action Workflow for instance allows the mapping of documental actions as the initiator of informational and even essential transactions.

What one observes when focusing on the pragmatic aspect of a business system in which actors carry on performing conversations : Actagenic conversations, resulting in agreements about future actions and factagenic conversations, resulting in the establishment of facts. Only in performing conversations, original new things are accomplished, we consider these conversations to represent the essence of the organisation. Consequently, we call the actions that are agreed upon in actagenic conversations and the results of which are established in factagenic conversations, essential actions, and the conceptualisation of the system observed the essential model of the organisation (Reijswoud, 1995).

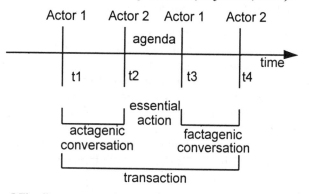

Fig. 5 The discrete transaction model

Every (essential) action is embedded in a transaction and every established fact is the result of the successful carrying through of a transaction (see figure 5). This constitutes the DEMO transaction paradigm. The transaction is the unit of analysis in organisational behaviour and is divided in three parts. In the actagenic conversation (e.g. negotiation phase) a commitment is stated that one of the parties (e.g. actor 2) will perform an essential action. The result of an actagenic conversation is called an agendum (singular of agenda). After execution of the essential action the third part of the transaction follows; the factagenic conversation. In this phase the actors are reaching an agreement over the outcome of the action. The result of the factagenic conversation is the creation of a fact, which states that the negotiated action has been performed. Actor 1 is called the initiator of the transaction and actor 2 the executor. In each of the time-intervals a breakdown can occur, resulting in an unsuccessful completion or a partial retry of the transaction. An extensive discussion of the transaction model in contained in (Reijswoud, 1995a).

DEMO starts with the set up of the communication model. The Communication model is the specification of the (elementary) actors and of the interaction and interstriction structure. By interaction structure is understood the mutual influence through being initiator or executor of transactions. By interstriction is understood the mutual influencing by means of created facts that play a role in the condition part of the behavioural rules that are executed in carrying through transactions. The Information model is the specification of the state space and the process space in the object world, i.e. all information which is produced and used in the transactions of the organisation. The Behaviour model is the specification of the behaviour rules of the actors (procedures for the execution of an essential action), i.e. the system dynamics. Actors in the model are restricted to the organisation, represented as one elementary actor, and its network. Depending on the required level of abstraction actors are decomposed. The execution of an essential action can depend on the outcomes of other transactions, such that transactions can lead to external purchasing or internal production transactions which have to be performed to complete the transaction.

The Communication Model (CM) can graphically be represented by means of a Communication Diagram. An actor is represented by a box. A Transaction type is represented by a disk. The operational interpretation of a disk is a store for the statuses through which the transactions of the type pass in the course of time. Because of the transient character of this information, the disk symbol is called a transaction channel. The facts that are created as a result of the successful carrying through of a transaction, are considered to be represented by true propositions stored in a fact bank. A bank is represented by true propositions stored in a fact bank. A bank is represented by a diamond. To show that the facts are the result of carrying through transactions of a particular type, the diamond symbol is drawn 'behind' the disk symbol of the transaction type. The combined symbol is too specific to represent internal fact types and boundary fact types. However, the actors in the system may also need to know external facts, i.e. facts created outside the system. The combined symbol is too specific to represent the external fact types, because the corresponding transaction types fall outside our scope of interest: they are not known. Therefore, the external fact types are arbitrarily put together in external banks, which are represented by only the diamond symbol. The actor who is the initiator of a transaction type is connected to the transaction channel by an initiator link. It is represented by a plain line with an arrow head at the side of the actor box, pointing to that box. Interstriction is represented by data links (dotted lines) between actors on the one side and fact banks and transaction channels on the other side. Figure 6 shows an example Communication Diagram.

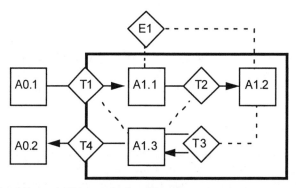

Fig. 6 Example Communication Diagram

The thick-lined rectangle represents the system boundary. The actors A0.1. and A0.2 are external actors, the actors A1.1., A1.2 and A1.3 are internal actors. The transaction types T1 and T4 are boundary transaction types, the transaction types T2 and T3 are internal transaction types. The bank E1 is an external bank. The external actor a0.1 is the initiator of transaction type T1; the internal actor A1.1 is the executor of this transaction type. In executing the transactions of this type and in carrying through the performing conversations, actor A1.1 uses data from the external bank E1. As a result of the execution of a transaction, it states one or more facts, which are stored in the bank of T1, and it initiates a transaction of type T4. Actor A1.3 is the executor of this transaction. In executing the transactions of type T4 and in carrying through he performing conversations, it uses data from the channel of transaction type T3 (transaction statuses). This transaction type has the same actor (A1.3) as initiator and as executor. We call this self-activation of actor A1.3 (Note, This is the way to model periodic activities). In executing the transactions of this type, and in carrying through the performing conversations, actor A1.3 uses elements of the contents of the bank of T1 and the channel of T3 as data. As a result of the execution of a transaction, it initiates a transaction of type T4. The executor of this transaction type is the external actor A0.2.

If information is understood to be the interdisciplinary field of science that deals with information and communication and their role in the functioning of dynamic systems, especially organisations, then the DEMO theory can most appropriately be called an informatical theory (Reijswoud, 1995a). In particular the theory does not draw on, and thus is independent of, any economic theory about organisational behaviour. This independence makes DEMO most valuable for providing an original informatical contribution to Reengineering the organisation, next to and in harmony with contributions from other disciplines.

3.2 Conclusion

In this article we have presented the Theory of DEMO as a fundamental new perspective for Reengineering Organisations through IT, which provides a Strategic as well as Informatical theory to Reengineering, this offers the following advantages:

- Modelling of different types of industries

  DEMO has an unique informational perspective on the organisations, which is not limited to one type of industry. Each transaction in DEMO consists of an essential action, such as delivering a product or service, and conversations, which allows information processing.

- Alignment of Business and IT strategy

  DEMO provides an ideal starting point for business Reengineering projects, because it distinguishes between essential actions on one side and informational and documental actions on the other side. This model of the essential characteristics leaves the right design freedom to the management and information experts. The transaction model in DEMO is based on knowledge how business systems operate. Thus redesign of transactions leads to specifications for systems redesign to realise Engineering Objectives.

- Decomposition

  DEMO abstracts completely from the organisational structure and technological way in which transactions are performed. This decomposition of transactions is intuitive and understandable by managers and employees and results in less paperwork compared to technical methods. The unique distinction between essential and informational actions in DEMO allows a scientific clean slate approach to determine the extent in which new technologies can be applied to obliterate old processes and to reengineer new transactions.

The successes of Business Redesign through Information Technology are well-publicised, but in practice organisations often have to rely on the intuition and experience of the management and IT-staff to realise the Reengineering objectives. DEMO provides a grounded theory, method and techniques which support the Reengineering of the organisation. Severally small and large projects have been performed in which DEMO has successfully been used as methodology for redesign and reengineering (cf. (Reijswoud and Rijst, 1994; Reijswoud and Rijst, 1995a; Reijswoud and Rijst, 1995b; Jansen and Poot, 1995)

References

Creemers M.R. (1993) *Transaction Engineering, process design and information technology beyond interchangeablility,* University of Amsterdam, Faculty of Economics and Econometry, Amsterdam.

Davenport Th. H. (1993) *Process Innovation, Reengineering Work through Information Technology,* Ernst & Young.

Davenport Th. H., and Stoddard D. B. (1994) Reengineering: Business Change of Mythic Proportions? *MIS Quarterly.*

Dewitz S. K., and Lee R. M. (1989) Legal Procedures as Formal Conversations: Contracting on a Performative Network. *Proceedings of International Conference on Information Systems,* Boston, pp. 53-65.

Dietz J. L. G., and Van Hee K. M. (1989) A framework for the conceptual Modeling of discrete dynamic systems. *Temporal Aspects in Information Systems,* Elsevier Science Publ. Comp.

Dietz J. L. G. (1991) The Essential System Model. *Lecture Notes in Computer Science 498,* Springer Verlag.

Dietz J. L. G. (1992) Leerboek Informatiekundige analyse (Dutch).

Dietz J. L. G. (1994) Business Modeling for Business Redesign, Proceedings of the Twenty-Seventh Annual Hawaii International Conference on Systems Sciences.

Dietz J. L. G. (1995) Denkwijzen, methoden en technieken (Dutch), HI-B2000, November 1995.

Flores F. M. and Ludlow, J. J. (1980) Doing and Speaking in the Office, in: Decision and Speaking in the Office, in: Decision Support Systems: Issues and Challenges, Pergamon, Oxford, 1980.

Hammer M. (1990) Reengineering Work: Don't Automate, Obliterate, *Harvard Business Review,* July-August 1990, pp104-112.

Jansen M., Poot F. C. J. (1995), *DEMO: a method for BPR (Dutch), een onderzoek naar de geschiktheid van DEMO voor het herontwerpen van bedrijfsprocessen: casus PTT Telecom,* TUDelft.

Keen P. G. W. (1992) *Shaping the Future, Business Design through Information Technology,* Harvard Business School Press.

Medina-Mora R., Winograd T., Flores R. and Flores F. (1992) The action workflow approach to workflow management technology. *Proceedings CSCW '92: Sharing Perspectives,* Association for Computing Machinery, New York, pp. 281-288.

Medina-Mora R., and Denning, P. J. (1995) Completing the Loops, *Institute for Operations Research and the Management Sciences,* pp. 42-57.

Porter M. E. (1980) *Competitive Strategy, Techniques for analyzing industries and competitors,* The Free Press, New York.

Porter M. E. (1981) The Contributions of Industrial Organisation to Strategic Management, *Academy of Management Review,* pp. 609-620 in: Barney, 1986.

Porter M. E. and Millar V.E. (1985) How Information gives you competitive Advantage, *Harvard Business Review,* July/August.

Porter M. E. (1985) *Competitive Advantage, Creating and sustaining superior performance,* The Free Press, New York.

Porter M. E. (1990) The Competitive Advantage of Nations, Harvard Business Review, March/April.

Van Reijswoud, V. E. and Van der Rijst, N.B.J. (1994) The Analysis of Planning Activities Using the DEMO Transaction Process Model. In: C. Lissoni, T. Richardson, R. Miles, T. Wood-Harper, N. Jayaratna (eds.) *Information System Methodologies 1994: Second Conference on Information Systems Methodologies of the Britisch Computer Society Information Systems Methodologies Specialist Group.* Heriot-Watt University Press, Edinburgh, pp. 283-294.

Van Reijswoud, V. E. and Van der Rijst, N. B. J. (1995a) Modeling Business Communication as a Foundation for Business Process Redesign : A Case of Production Logistics. *Proceedings of the 28th Annual Hawaii International Conference on System Sciences,* IEEE Computer Society Press, Los Alamitos, CA, pp. 841-850.

Van Reijswoud, V. E. and Van der Rijst, N. B. J. (1995b) Modeling Business Communication for the Purpose of Business of Process Reengineering. In: N. Jayaratna, R. Miles, Y. Merali, S. Probert (eds.) Information Systems Methodologies 1995: *Third Conference on Information Systems Methodologies of the Britisch Computer Society Information Systems Methodologies Specialist Group.*, pp. 173-184.

Scott Morton M. S. (1991) *The Corporation of the 1990's: Information technology and organizational transformation,* Sloan School of Management, Oxford University Press, New York.

Winograd T., Flores F. M., (1986) *Understanding Computers and Cognition,* A New Foundation for Design, Ablex Publishing Corporation, Norwood, New York.

Yourdon E. (1989) *Modern Structured Analysis,* Prentice–Hall International.

Part 4   Business Process Modelling - Experiences

Information Technology as an Enabler of Business Processes Designing During Macroeconomic Transformation

M. A. Janson, St. Wrycza

Integration of IT Strategy, Process Analysis and System Implementation in an Office Automation Based Process Improvement Programme

D. J. Chaffey

Information Technology as an Enabler of Business Processes Designing During Macroeconomic Transformation

Marius A. Janson[1] and Stanislaw Wrycza[2]

[1]Department of Information Systems, University of Missouri-St. Louis, 8001 Natural Bridge Road, St. Louis, MO 63121, U.S.A., mjanson@umslvma.umsl.edu
[2]Department of Information Systems, University of Gdansk, Ul. Armii Krajowej 119/121, 81-824 Sopot, Poland, ekosw@halina.univ.gda.pl

Abstract. The influence of information technology (IT) and market opportunities on business processes reengineering (BPR) and later designing in small firms in course of macroeconomic transformation is explored and reported in this paper. The exploratory, qualitative research method was used to analyse and clarify a data structure. Seven small Polish firms were selected for the survey as a convenience sample. The companies show the use of technology ranging from the standard to advanced one. In the result of research process, the model of new business processes design is proposed. It is based on the idea of the combined IT initiative/Market demand BPR enabler. The model was constructed and then verified by means of exploring, analysing and generalising data acquired in the interviewed companies, and synthesised in tabular form in the paper.

Keywords. information technology (IT), business processes reengineering (BPR), small firm, macroeconomic transformation, BPR enabler, exploratory qualitative research method.

1 Introduction

Information technology (IT) is an essential component of a firm's strategy in a global market (Ives, et al., 1993). Hence, it is not surprising that academics and practitioners have focused on identifying key IT issues, first in Western Industrialised Countries (Dickson, et al., 1984; Hirschheim, 1988; Niederman, et al., 1991; Watson, 1989; Watson and Brancheau, 1991) world and subsequently also in Eastern European countries (Dekleva, and Zupancic, 1992; Wrycza, and Plata-Przechlewski, 1994). These studies uncovered a broad range of IT concerns which can be classified into two main categories: technological and managerial concerns. Dexter, et al. (1993) found that most pressing managerial IT concerns in Estonia were technical whereas in Western industrial countries managerial IT issues dominated.

The above mentioned surveys described the views, problems, opinions, and threatens of Central European managers and IT professionals at the beginning of macroeconomic transformation in this region. The Central European countries are now rather at the end of transition process, functioning mostly in the conditions of market economy. This situation evokes new questions. What is the influence of the information technology on business expansion, especially in a new private sector? What is the level of the acceptance and adoption of the current important and prospective IT solutions and initiatives in this region? Does the IT stimulate to start the new business processes and firms? What type? Or do the Central Europeans give their own original contribution to IT/IS modern development? It seems that the above and similar questions need more in-depth analysis. Therefore we have applied the exploratory qualitative interview-based method to answer the questions and explore the strategic IT/business liaison in this region. We have selected seven small firms in Poland - the country with the longest economic transformation and market economy experience in Central Europe.

The paper starts with Introduction. In the second chapter the principles of the selected exploratory, qualitative, interview-based research method as well as the convenience sample are described. Next, the evolution of small business sector in Poland is presented. The main findings of the research concerning the relationships between IT and business processes enabling and designing as well as their model are comprised in part four. The paper is finalised with concluding remarks.

2 Research Method and Convenience Sample

Our project is a qualitative interview-based attempt at uncovering and making explicit the conditions and individual actions surrounding the creation and development of entrepreneurial initiatives in Poland. Hartwick et al. (1994) elaborated the different objectives of confirmatory and exploratory research approaches. Confirmatory research seeks support for hypotheses which have been formulated using an a priori theory. These hypotheses are subsequently tested using data from carefully controlled experimental studies.

Exploratory research proceeds differently than confirmatory studies - rather than relying on large randomly selected samples, exploratory analysis employs small convenience samples. Instead of using test statistics based on assumed probability distributions, exploratory analysis for the most part uses tabular and graphical data displays and other techniques for probing the data flexibly (McNeil, 1977). Exploratory analysis represents a first contact with the data and aims to isolate data patterns and features, and to uncover, clarify, and simplify a data structure (Velleman et al., 1981). The data acquire meaning within the analyst's frame of reference, because he or she searches for a satisfactory structure by confronting the data with a variety of alternative models (which is to say: the model follows the data). In this spirit exploratory data analysis is defined by Benzecri (1980) as a "method that extracts structure from data".

Before scheduling a series of interviews with entrepreneurial talents in Poland we first fixed our project's goals. The semi-structured interview sought information on the: 1) age of the business and changing business conditions resulting from the demise of ancient regime, 2) number of partners and type of company's managerial structure, 3) number of employees and their contribution to achieving company objectives, 4) nature of the business and competition, and current and future markets, 5) business clients and the services delivered, 6) company's operating philosophy, 7) the company's use of IT to: a) serve the client, b) manage and run the company, c) manage and maintain the company's information systems, and d) the company's use and acquisition of standard as well as special software and hardware (e.g., wordprocessing, spreadsheet, and data base), and 8) the balance between providing clients a quality product or service, workers meaningful work, realising adequate returns on investment, and staying ahead of the competition. The detailed descriptions of the survey and the extensive tabular comparisons of the results are presented in Janson and Wrycza (1996).

We interviewed seven small business firms in Gdansk area, selected randomly. Gdansk agglomeration is one of the fastest economically growing regions in Poland. The excerpt of the survey is shown in Table 1 (particularly quantitative data). All interviews except one, were conducted in English. Despite their stark exteriors office buildings were always pleasantly arranged inside. All seven companies had made considerable investments in personal computers, office furniture, stationary, binders and boxes. Employees were professional, polite, patient and did not interrupt. The majority of employees understand and speak English well. Furthermore, employees displayed team spirit which, we suspect, arose from a flat hierarchical management structure.

Employees had great freedom to arrange their own activities and they shared the professionalism of their employers ; they took care of their outward appearance, dress and they were dedicated to their organisation. Five of the seven use English names. The two exceptions are organisations which were active before 1989 and have Polish names "Doradca" (Advisor) and "Kancelaria Notarialna" (Notary Public).

No.	Firm	Business Start	Number of partners	Number of employees	Number or type of clients	Business nature
1.	English Language School	1989	4	15	1000/year	• High quality English Education
2.	Young Digital Poland	1989	4	11 (high-tech education and experience)	over 1000	• Acoustical Instrumentation • Multimedia Software
3.	Doradca	1985	15	30	State-owned firms, co-operatives, independent firms	• Consulting services, • Privatisation, • Financial Marketing, • Valuation
4.	Combi Data	1992	3	15-20	150 students, City Hall, Emergency Management	• End-user education, • GIS Systems, • Emergency Information System
5.	Kancelaria Notarialna	till 1989 - state-owned 1991- reprivatized	3	6	Individuals, Companies	Legal Services
6.	DC	1989	3	28	Individuals, Companies	• Software and Computer Selling, • University Software Development, • Insurance Software, • End-User Education

No.	Firm	Business Start	Number of partners	Number of employees	Number or type of clients	Business nature
7.	Cross/ Comm	1991	2	150	150 Companies	• Routers Software Development, • Networks (LANs and WANs) installation and services

Table 1: The Convenience Sample Characteristics

By using an English name the company makes explicit its intention to be active internationally. One interviewee explained that firms prefer English names because it makes it easier for their international partners and customers who would have a hard time pronouncing the strings of consonants that occur in many Polish names. Not everyone is pleased that Polish firms have English names and the practice is at times criticised in the media.

3 Small Business Development In Poland

Prosperity in and political independence of Poland depend on the nation's successful transformation from a command to a free market economy. Economic transformation imposes excruciatingly difficult economic conditions on the populace. The future of Poland is far from certain and if economic transformation were to stall the nation may be pushed into the grey sphere of economic stagnation. Information is a key to successful economic transformation.

During the interwar (1918-1939) years private companies in Poland were the rule rather than the exception. In the 1940s nationalisation of factories, real estate, and land commenced. However, small farms were allowed to exist as well as some small shops and handicrafts. During the 1980s there was some growth in the numbers of small firms that provided services. These were mostly organisations at the periphery of the national economy such as, for example, small farmers, shops, handicrafts, or consulting firms such as Doradca. With the demise of the ancient regime in 1989 the private sector has experienced substantial growth. As shown in Table 2, the percentage of employment has shifted away from the public toward the private sector. Whereas in 1990 the private sector counted for less than 50% of the nations employment, in 1994 this figure had increased to over 60%.

Despite their precarious legal status businesses that survived the period before 1989 started the process of economic transformation at a surer footing than firms which were formed during the early 1990s.

212

Year	Employment (Thousand)	Public Sector (Percent)	Private Sector (Percent)
1990	16,145	51.1	48.9
1991	15,443	45.7	54.3
1992	15,011	44.0	56.0
1993	14,761	41.1	58.9
1994	14,923	39.0	61.0

*Polish Small Statistical Yearbook, GUS, Warsaw 1995, p. 329

Table 2: Employment in the Public and Private Business Sectors in Poland

The older organisations were better prepared legally and had more business experience upon which to build future growth. Also, during their infancy the pre 1990's firms had fewer competitors and did not have to invest immediately large sums of money in information technology. Hence, these older firms were better prepared than their younger counterparts to face the harsh economic realities such as the decrepit telecommunications system. In fact, even today the way around telecommunication's problem is to sign up on the INTERNET or to acquire bandwidth on a satellite service. This is precisely what two of the seven companies have done: Young Digital Poland is on the Internet while Cross Comm has access to a satellite system. However, these are high cost solutions to problems brought about by a poor telecommunication infrastructure which few companies can afford.

4 Business Process Reengineering or Business Process Inventing

BPR - Business Processes Reengineering focuses the attention of many researchers and practitioners in the field of information systems development. The investigation carried out by us confirms that BPR is extensively practised in Polish firms being in the process of transformation. Hammer and Champy (1993) define reengineering as "the fundamental rethinking and radical redesign of business processes to achieve dramatic improvements in critical contemporary measures of performance such as cost, quality, service and speed". Yes, the macroeconomic transformation which took place in Poland and later in the other Central and Eastern European countries at the beginning of nineties was fundamental, radical, dramatic and critical.

It created a wide free space for individual and corporate activities for people with the entrepreneurial spirit and readiness to accept the tough rules of free market economy. As a matter of fact, it was the free space needed for BPR accomplishment in the business, microeconomic scale. Business processes were enforced or rather introduced in a natural way (by the way - without knowing the BPR theory) by global transformation of economic system.

There have been two spheres of BPR application:
a) in the public sector, in big firms, employing several thousands workers it was classical BPR just to survive, reconstruct, compete effectively and finally expand at the market,
b) in the private sector, practically in small business, in over three million new companies, which started form the scratch - it was what we call business processes inventing or at least initiating.

These before non-existent firms have had to find their way to function in free market economic conditions.

The research concentrated around the second sphere taking into account its uniqueness. It is, of course, an extreme example but compliant with the general BPR definition. In particular we have tried to explore the influence of information technology and systems on new business processes design and later on the performance in small firms. Therefore, according the exploratory research method where model follows data we probed data flexibly. and confronted them with different alternative models. The closest description was connected with the Davenport's (1993) model:

IT Initiative--------->Process Change---------->Economic Outcome

It proved very quickly that PC technology is not the IT stimulus or innovator we are looking for. Practically all small firms, starting from a one-man travel office are equipped with PCs, and peripherals.

Staff is effective in using and operating PC systems, wordprocessors, spreadsheets, databases and bookkeeping systems. They are simply a "must", a standard for small business. An advantage of starting late?

Analysing and comparing different cases we came into some general regularities observed in the successful firms. It seems that their IT-Process-Productivity relationship is like on the figure 1. There are two process enablers which are strictly connected and act parallel. The combination of IT technology and the market niche is the initiator of the business processes establishment and development. They interact in a new process initiating and designing. A newly designed process means very often also a new small company. Two initiators of the new process invention play the role of "technology push" and "market pull" respectively. A country's economic and technological level as well as the foreign trade challenges determine the opportunities and threats for newly invented and implemented ideas.

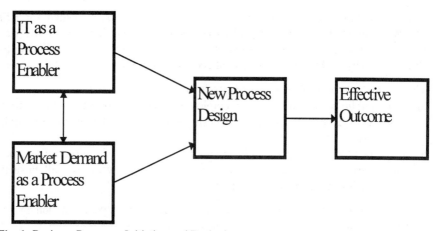

Fig. 1: Business Processes Initiating and Designing

There are technological achievements not accepted by the market which maybe too narrow or not prepared. The cases of bankruptcy or stagnation are still frequent. However, if the firm or the business process is successful than the final positive outcome can be considered in the respect of firm development and in a broader macroeconomic context advantage.

Our findings in using the paradigm (figure 1) for exploring the problem with respect of the selected small business firms are synthesised in table 3.

When looking at the IT initiatives for starting the process invention and designing it is visible that very up-to-date solutions and technologies are used, for example satellite transmission, multimedia technology, different sorts of computer networking, and business applications of INTERNET. On the other hand, market opportunities imply services and production for foreign partners, resulting on reduction of underinvestments and delays in certain sectors of business and social life, by coping with the challenges of new technological developments and cultural standards. There is a whole range of visions and than designs of new processes invoked by the two basic, combined stimuli. They include modern private education on different levels, development of multimedia and network software, information systems for emergency managing. Some of the firms concentrate only on one process, while others are taking on new opportunities, developing several streams of versatile processes. For example CombiData except emergency management analysed in table 3, develops geographical information systems using SUN workstations, runs a professional school for computer network specialists, and offers outsourcing for management information systems. The effective outcomes are products and services strictly connected with the innovative processes. It is worth to stress that managerial

No.	Name of the firm	IT Enabler (Technology Push)	Economic or Social Market Demand Enabler (Market Pull)	New Process Design	Effective Outcome
1.	English Language Services	CD ROM, Interactive Video	Opening of foreign trade (state - licensed till 90ties) to all - public and private companies. Drawbacks and delays hitherto educational system - very limited opportunities to learn Western languages Establishment and daily expansion of numerous international business ties	Modern, effective, competitive teaching of businessmen and managers aiming at or involved in foreign trade activities with English grades recognised internationally	High quality education business in using modern method technologies and staff (Polish, British) Alumni with the recognised worldwide grades like FCE or Proficiency of Oxford University The management of many small firms able to communicate, operate and compete on foreign markets
2.	Young Digital Poland	Multimedia technology, Business applications of the Internet	The growing significance of multimedia technology and its share of educational aids market in Western industrialised countries Well-skilled Polish professionals in multimedia technology	Development of multi-media software and multimedia products, marketing and selling them on the markets of Western industrialised countries	Multimedia software and CD-ROM products Multimedia laboratory for product development Management and staff coping with up-to-date technology, developing it and competing on foreign advanced markets
3.	Doradca	LANs, Accounting information systems	Shift and adoption into Western style of accounting in both public and private sectors of economy Privatisation and reprevatisation globally in economy and firms Privatisation expertise needed in the other countries in transformation, for example in Russia	Consultation and expertise in up-to-date accountancy and privatisation of business Implementation of the Western-style accounting information systems Accomplishment of company's privatisation	Widespread acceptance of the Western - style accountancy and its implementation in business organisations (in a computerised version) Privatisation of firms in Poland and abroad
4.	Combi Data	Metropolitan Area Networks, Cellular Phones	The social demand for quicker and more reliable emergency actions in case of accidents	Managing city emergencies involving police, hospitals, city officers	The effective, reliable emergency information system
5.	Kancelaria Notarialna	LANs, Databases, Legal Information Systems	The growing number of legal transactions and operations particularity in the private sector	Computerisation of legal operations and its databases	Information system supporting the daily operations of the legal notary
6.	DC	Software generators, CASE tools	Improvement of university administration functioning New, additional tasks for university administration	Restructuring the University administration procedures Software development to support university administration	Management information system for university administration, unified for all universities

No.	Name of the firm	IT Enabler (Technology Push)	Economic or Social Market Demand Enabler (Market Pull)	New Process Design	Effective Outcome
7.	Cross/ Comm	Computer Networks, Satellite transmission, Acoustical instruments	The increasing demand for WAN and LANs in different businesses and institutions of the country	Computer networks installation, services and programming	Reliable provider of advanced computer network services compliant with ISO 9001 certification Network software for routers

Table 3: New Processes Design

focus of new firms is on high technical or quality standards. This network software is compliant with ISO 9001 standards, accountancy with Western-style, language courses with Oxford University schemes. The success of these outcomes is the verification of the correctness of the new processes design.

5 Conclusions

The qualitative, explorative method, used in this paper, allowed to explore and then construct the model of business processes design in small firms, starting to operate in the conditions of macroeconomic transformation to free market economy. The empirical data, which extracted into model proved that combination of modern IT initiatives and market opportunities is the basic enabler of the business processes reengineering (BPR) or designing. The data structure was indicated by details of the other elements of the model, i.e. new processes design and effective outcome, gathered in convenience sample and presented in Table 3. But besides the model there is a human side of business processes initiating and development in economically transforming countries of Central Europe like Poland. Transforming opportunities into business success requires managers with the insight, flexibility, and decisiveness of the seven partners discussed by our study.

References

Benzecri, F. (1980) Data Analysis, Paris, France: Dunod

Chalmers, J. (1993) "Survey Telecommunications," Business Central Europe, Vol.1, No.1, 35-48.

Cook, J. (1993) "Fighting Blue Beards: Intellectual Property Pirates Still Roam the Czech Seas," Business Central Europe, Vol.3, No.20, 60.

Davenport D. H. (1993) Process Innovation: Reengineering Work Through Information Technology, Harvard Business School Press, Boston

Dekleva, S., and Zupancic, J. (1992) "An Empirical Investigation of Key Issues in Information Systems management," in Proceedings of the Third International Conference, Poland: Sopot, Wrycza (ed.), 95-104.

217

Dexter, A., Janson, M., Kiudorf, E., and Laast-Laas, J. (1993) "Key Information Technology Issues in Estonia," Journal of Strategic Information Systems, Vol.2, No.2, 139-152.

Dickson, G., Leitheiser, R., Nechis, M, and Wetherbe, J. (1984) "Key Information Systems Issues For The 1980s," Management Information Systems Quarterly, Vol.8, No.3, 135-148.

Frost and Sullivan (July/August, 1995) "Information Technology Special Supplement: Visegrad Realities," Business Central Europe, Vol.3, No.23, III XVIII.

Hammer M., Champy J. (1993) Reengineering the Corporation, Harper Business, New York.

Hartwick, J., and Barki, H. (December,1994) "Hypothesis Testing and Hypothesis Generating Research: An Example from the User Participation Literature," Information Systems Research, Vol.5, No.4, 446-449.

Hirschheim, R. (1988) "An Exploration Into The Management Of The Information Systems Function: Key Issues And An Evolutionary Model," Proceedings of the Conference On Information Technology Management For Productivity and Strategic Advantage, Singapore, 4.15-4.38.

Ives, B., Jarvenpaa, S., and Mas, R. (1993) "Global Business Drivers: Aligning Information Technology To Global Business Strategy," International Business Machine Journal, Vol. 32, No.1, 143-161.

Janson, M., Wrycza S. (1996) IT in Small Companies: A Case Study Of Business Transformation in Poland (to be published).

McNeil, D. R. (1977) Interactive Data Analysis: A Practical Primer, New York, NY: Wiley.

Niederman, F., Brancheau, J., and Wetherbe, J. (1991) "Information Systems Management Issues in the 1990s," Management Information Systems Quarterly, Vol.15, No.4, 475-500.

Pullar-Strecker, T., and Papp, B. (December 1993/January 1994) "Manager Wanted," Business Central Europe, Vol.1, No.7, 7-9.

Velleman, P., and Hoaglin, D. C. (1981) Applications, Basics, and Computing of Exploratory Data Analysis, Boston, MA: Duxbury Press.

Watson, R. T., (August, 1989) "Key Issues in Information Systems Management: An Australian Perspective - 1988," Australian Computer Journal, Vol.21, No.3, 118-129.

Watson, R. T., and Brancheau, J. C. (March, 1991) "Key Issues in Information Systems Management: An International Perspective," Information and Management, Vol.20, No.3, 213-223.

Wrycza, S., and Plata-Przechlewski, T. (1994) "Key Issues in Information Systems Management: The Case of Poland," in Proceedings of the Fourth International Conference on Information Systems Development, J. Zupancic, and S. Wrycza, (eds.), Bled, Slovenia, 289-296.

Integration of IT strategy, process analysis and system implementation in an office automation based process improvement programme

Dave J. Chaffey

School of Business, University of Derby, Kedleston Road, Derby DE22 1GB, United Kingdom, d.chaffey@dertby.ac.uk

Abstract. The strategic options available to a business wishing to implement workflow automation of existing manual processes are examined through a case study. To implement the business strategy a process re-engineering exercise can be undertaken, or a less radical process improvement can occur. It is shown for a national estate agency (real estate) business in the UK why this decision was taken. A phased method of implementation based on process improvement rather than re-engineering was selected since it was thought to represent the smallest commercial and technical risk and would gain the greatest support from the staff. Identification of key processes based on Critical Success Factors (CSFs) ensured that the business processes encapsulated in the IT system were consistent with the strategic objectives. A hierarchical, activity based method of process analysis was undertaken which encompassed both the identification of business processes and the process elements forming the basis of a Workflow Management (WFM) component of the customer facing application. It is shown that to fully model the 170 tasks performed in this business, a four level nested classification was more appropriate than a simpler two level approach. The higher level abstraction avoids the lower level complexities of the business rules and operations thus facilitating understanding of the process analysis by the business users. The enactment of the Level 2 activities in the WFM system was developed using a workflow scheduler based on standard Oracle database triggers. These were used to automatically generate tasks following state changes of objects within the system.

Keywords. Process analysis, process improvement, office automation, workflow, alignment of new processes with business strategy

1 Introduction

A company which wishes to use Information Technology to gain competitive advantage is today faced with a choice of whether to base the introduction of technology on automation of the existing business processes or to re-engineer the business with automation of substantially different processes (Davenport, 1993). With both approaches Workflow Management (WFM) software can be leveraged to assist in the automation of these processes. This paper considers the case history of a UK based estate agency faced with this choice. The previous IT facilities were limited to a single PC or typewriter in each branch for a secretary to prepare property details. The background to the new IT strategy is described for this business area by considering the factors driving the strategy and the choice of whether to re-engineer or simply automate the process. The stages involved in implementing the strategy are considered with an emphasis on how the business processes were modelled and then transformed into an application intended to fulfill the strategic objectives.

2 Business strategy, the background to process improvement

All players in the UK property market face intense competition from several national agencies and some of the many local agents. The company in this case study, a national agent required a strategy that could apply technology to gain a market advantage. The principal objective was to increase market share nationally while retaining the same number of branches at existing staff levels. The increased market share was to be achieved through using a computer system as a differentiator from competitors to attract customers wishing to sell their properties. This market differentiation was to be achieved by producing a highly graphical, customer facing application with the main interactive screen consisting of a location map indicating the position and details of properties including photographs. It was intended that this would increase property listings due to property vendors believing that more applicants would be attracted to their property. To meet the hoped for increase in volume of business it was thought that WFM software could be used to partially automate the customer service and administrative functions while improving service quality. The main target users of the system (*actors* (Curtis et al, 1992) or *agents* (Georgakoupoulos et al, 1995) within WFM terminology) were *sales negotiators* who arrange the sale and purchase of properties, *managers* who control the performance of a branches business and *branch administrators* who prepare details of properties and dispatch routine correspondence to customers. The workflow component was intended to improve customer service through prompting negotiators to contact customers more regularly and ensuring that all contacts were recorded. Several improvements in data quality were anticipated, firstly to record a full status history for each property and applicant as a transaction log detailing each customer interaction. Secondly, it was hoped to increase the capture rate and accuracy of information concerning customers and properties. Finally the system was designed to provide asynchronous communication between staff to improve co-operation between negotiators who are often out of the office.

The main objectives of the project are summarised in Fig. 1 which categorises benefits according to how easily they are measured. It is evident that although a range of benefits were identifiable, it is less easy to assess whether they are achieved and the IT strategy fulfilled.

		High	• Increase listings • Maintain staff levels with increased volume of work • Reduce assets	• Improve data quality • Increase market share
Tangible		Low	• Faster access to information	• Improve customer service • Improved sharing of information between branches
			High	Low
			Measurable	

Fig. 1: Categorisation of the benefits of implementing an IT system in the estate agency business

3 Strategy implementation

An early decision required in the implementation of the business strategy was whether to adopt and improve existing processes in a process improvement programme or to re-engineer them. Three basic alternatives were available to the company to implement the business strategy described above, these are similar to the different alternatives proposed for "Process Innovation" by Davenport (1993) and Tsalgatidou and Junginger (1995). The alternatives were:

- to automate the existing working methods using computers to assist in performing tasks, but retaining the existing task structure
- to base the office automation on the existing processes, but identify key areas for process improvement such as the most time consuming tasks
- to re-engineer the business and make major modifications to the processes and the roles of the agents performing them

3.1 Taking the re-engineering versus improvement decision

The intermediate option was chosen since it was thought that this could yield significant improvements in staff productivity, but without the higher risk of re-engineering. The failure rate of re-engineering exercises is estimated as high as 90% for projects that either do not achieve their strategic aims or overrun their budget(*Computing* magazine, *2/11/95*), Davenport (1993). Thus although re-engineering was thought to give potentially the largest gain, it also poses the largest risk, particularly since no IT infrastructure existed before the current project. Many companies who perform re-engineering will previously have had the benefit of implementing an IT system to partially automate the business process. Through this they will be able to learn from mistakes and identify processes which will particularly benefit from re-engineering (Hammer and Champy, 1993). It was thought by the project management team in this example that process improvement, as an incremental approach would be most appropriate for gradual introduction of technology to end users who had limited previous experience of using computers. A rapid introduction of the system based on application prototyping with a high degree of user involvement was intended in order that several iterations of the application could be made within a period of a year, before further rollout. Once the new system was in place and evaluation complete it would still be possible to identify further improvements for later stages in the project. A final reason for choosing a gradual approach was the rapid evolution in technology such as internetworking facilities and WFM which once the technology matured were thought to offer a completely different paradigm for developing the application. In the future the system could either be accessed from home using the World Wide Web, or from customer information kiosks deployed in shopping malls.

The phases envisaged in the project to fulfill the strategy are summarised below and described in more detail in the following sections:
1. Preliminary review of whether to opt for improvement or re-engineering
2. Analyse and model existing business processes in detail (3 months) in one area - the test-zone environment. Identify key processes for improvement
3. Develop application through RAD and prototyping (4 months)
4. Deploy application in single branch (1 month)
5. Review application in single branch and assess WFM volume (1 month)
6. Revise application (1 month)
7. Roll-out to 10 branches in test zone (2 months)
8. Review implementation through work measurement, customer
9. and negotiator questionnaires
10.Install new version(s) as appropriate in further incremental improvement
11.before national rollout (9 months).

3.2 User requirements

The functional requirements of the application were driven by the system being used directly with customers. Owing to the need to capture the customers interest and attention and then maintain it the following were important requirements:

- good usability through a simple user interface with high graphical content giving easy access to maps of property location, photographs and customer information
- seamless integration of workflow and image retrieval into the system
- acceptable information retrieval and display performance

3.3 System architecture

The system architecture in the test zone consisted of 10 branches and an area office connected using a Wide Area Network (WAN). Data was accessed using a simple client/server arrangement with PC clients accessing data from a central Windows NT hosted Oracle database server across the WAN. Details of the infrastucture ares described in Chaffey (1996) which discusses the factors involved in implementing such a system across a WAN. It was hoped that market share could be increased by making the branch network representation more effective through linking each branch via a WAN to a central branch database server holding details of all properties.

4 Process analysis

Process analysis was conducted using a traditional system analysis approach in contrast to the synthesis approach described by Kariagiannis (1995). The analysis approach based on comprehensive capture of processes and a detailed task representation was found to be suitable for an implementation which was intended as a process automation and improvement exercise rather than a re-engineering exercise. With a re-engineering project it is likely that the synthesis method would be better since it leads to more alternatives for re-engineering. Analysis consisted of the following stages:

1. Information collection (knowledge acquisition)
2. Process modelling to identify the main business processes, coupled with data modelling based on the Entity-Relationship (ER) approach to assist in database construction. Gruhn and Wolf (1995) also describe a case study where these two techniques are used together.
3. Identification of detailed activities and tasks forming the main processes. This stage also involved identification of key areas of negotiator work where the

greatest potential process improvement could occur to assist in achieving the CSFs.

4. Evaluation of suitability of processes as part of prototyping the system.

These stages are similar to those of, for example Tsalgatidou and Junginger (1995) and are described in more detail below.

4.1 Information collection

The plan for information collection was to make good use of existing process descriptions, but to supplement these with a programme of observation in branches and negotiator questionnaires. Sources defining existing processes included:

- Written customer service standards for standard branch operations such as processing a new offer on a property
- Forms for obtaining customer and property details
- Notebooks used in the office for sharing information such as an office diary and a log of all viewings that had occurred
- Standard letters sent to customers on key events such as following registration
- A paper based "Branch Information System" - a log book containing branch performance indicators each month in terms of number of sales and financial performance

From analysing these information sources, observation and interview it was possible to identify all the different process elements in the business that are referred to below.

4.2 Identification of the main business processes

This was achieved through considering the main Critical Success Factors which were defined as the factors necessary for a branch to be profitable, based mainly on good customer service and efficient business processes.

The Critical Success Factors and how they would be achieved through their associated processes are shown in Table 1.

CSF	Achieve through	IT system to assist through	Business process
Achieve inspections	Attracting vendor for valuation of property	Illustrating efficiency of business (workflow, status indicators). Larger number of applicants will see property	Obtain listing

CSF	Achieve through	IT system to assist through	Business process
Convert inspections to listings	Competitive commission and sales skills	Arguments as above	Obtain listing
Market property to ensure offers received and accepted	Attracting applicant to branch, viewing properties	WFM component to prompt negotiators to perform marketing activities and update vendors on status	Market Property
Achieve completion revenue	Controlling progess to completion frequently	WFM component to prompt negotiators to check with all parties involved in sale	Facilitate sale
Generate mortgage revenue	Financial advisors provide quotation	System use by Financial advisors not seen as a priority	Financial services
Performance of sales staff	Personal characteristics	System to be used as a sales tool	All above processes
Managers to ensure that CSFs are achieved	Business & management skills	Monitoring of staff work-flow Monitoring of branch performance by reports	Manage Process

Table 1: Critical Success Factors for the estate agency business and how they are achieved through business processes

4.3 Identification of detailed process elements

A four level process model was used to describe both the principal business operations and their constituent *process elements* (Curtis et al, 1992) performed by staff. This model used an activity based method (Marshak, 1994) rather than a communication based methodology or customer supplier protocol (Scherr, 1992, Teufel and Teufel, 1992). This approach was thought to be appropriate to the real estate business because of the relatively unstructured, non-sequential method of working caused by the need to react to phone calls and customer visits to the branch. The customer-supplier protocol is focused on agreed customer satisfaction levels for each process. For the large number of tasks performed each hour it is not practical to establish detailed quality criteria for each, other than general criteria that the task is completed professionally, accurately and as soon as possible. A work measurement analysis in four branches over a five day period highlighted these characteristics of customer-negotiator interactions:

- non continuity of tasks caused by customers entering the branch or calling (average of 7 calls, 1 personal contact and 2 conversations with other negotiators per hour)
- the high proportion of time spent on the telephone (average of 3.7 calls in and 3.3 calls out per hour totaling 13 mins per hour)
- the large number of separate tasks performed (an average of 28 different tasks per hour)
- limited incidence of some tasks which occur infrequently such as registering a vendor in the branch.

When defining the processes the team also tried to identify factors affecting the quality of service for each process. Quality in the service sector can be measured by the Service quality gap method SERVQUAL (Parasuraman et al 1998) which is based on customer perception of expectation of service compared to actual service obtained through interviews. It was found not to be appropriate to use such a method for the hundreds of tasks which were identified for this business area.

5 Process definition method

Processes were identified in a four level hierarchy similar to the nested task approach as described for example in Georgakoupoulos et al (1995). Each process level consists of *process units* or *elements* (Curtis et al, 1992) as follows: each level 1 *business process* is decomposed into Level 2 *activities* which are further divided into level 3 *tasks* and level 4 *sub-tasks*. Each process element such as a level 2 activity has *actor(s)* e.g. Curtis et al (1992) which perform the process, *object(s)* such as customer and or property which the process acts on and a *conclusion* which is a state transition of the object when the process element is complete (an *event* recorded in the WFM) together with initiation of a subsequent process.

Level 1: **Business Process** - one of the main business processes. Defining processes at this level is important to ensure that the tasks lower in the hierarchy correspond to the business objectives of Table 1. For the real estate business the main processes involved with selling a house are *obtain listing* of property, *market property*, *facilitate sale* and *provide financial services*. Davenport (1993), notes that even for large multi-national organisations the number of main processes will rarely exceed 10.

(Number of level 1 processes = 5)

Level 2. **Activity** - a significant negotiator operation to partially fulfill a particular business process and usually associated with an object state transition (see below). The activity is recognised by managers and negotiators as one of the main stages in house sale for example "Arrange viewing" for a potential purchaser. The number of process elements and their description at this level is such that end-users can easily understand them and complete them in a few minutes. This level of detail was that typically used in the WFM task list.

(Number of level 2 activities = 34)

Level 3. **Task-** a piece of work which is part of the activity that can usually be completed immediately, but can be broken down further. Some tasks in the workflow task list were specified at this level, but there were too many in total to be specified.

(Number of level 3 tasks = 136)

Level 4: **Sub-task-** the smallest division of work used in work measurement. This would be at the level of "walk to filing cabinet, retrieve card, return to desk". Clearly it was not practical to specify tasks in the workflow task list at this level.

It is more usual in the literature to adopt a two (or sometimes three) tier hierarchy with *business processes* at the top level and *activities* (Tsalgatidou, 1995,) or *tasks* (Georgakopoulos et al, 1995, Rusinkiewicz, 1995) at the next level. Owing to the number of tasks in a real business, decomposition of the process into more than two tiers was necessary to clearly represent the processes. With a simple two level hierarchy there would be five processes at level one and more than 170 process elements at level 2. Tasks in the WFM component were represented at level 2 since the degree of detail for level 2 with approximately 40 tasks was practical for this.

Following information collection, the four level process model was produced by a top-down decomposition into process elements. An example of the 4 process levels for the *Obtain listing* business process is given as Figure 2. The types of tasks recorded at level 3 are detailed for the level 2 activity *Ensure listing obtained.* A pictorial representation for each level 3 task was adopted to assist system analysis by indicating whether information capture or retrieval was to occur and whether a reminder was required for the negotiator to perform the task. Following completion of the system it was found that 85% process coverage occurred for sales activities performed by negotiators at level 2 i.e. 85% of activities were conducted using the system. Other activities tended to be those that occurred out of the office.

Level 1 Business process:

Obtain listings

    Level 2. Activities:

    🖥 *1. Register vendor*

    🖥 *2. Inspection preparation*

    🖥 *3. Inspection appointment*

    🖥 *4a. Follow up Inspection - IF listing instructions obtained*

    🖥 *4b. Follow up Inspection - IF listing instructions NOT obtained (exception)*

    🖥 *5. Ensure listing obtained*

        Level 3. Tasks for "ensure listing obtained"

        📁✒   *1. Prepare initial property particulars (48 hours)*

        🕐   *2. Remind negotiator to handle sale to contact vendor*

        ✉   *3. Send property particulars*
        ✉   *4. Instruction to sell with draft PMA details*

        🕐📁✒ *5. Contact vendor to obtain PMA on particulars*
        Level 4. Sub-tasks:

            ✉   *1. PMA chase*
            ✉   *2. Tenure details*
            ✉ *3. Amended details*

Key

🕐   Reminder required for negotiator

✒   Record Information (data input)

📁   Retrieve Information (data output)

🖥   IT system used to perform activity

✉   Letter or form produced (level 4)

Fig. 2: An example of the 4 level process analysis for the real estate business. Note an Inspection is where the property to be sold is visited by the negotiator to assess its value

5.1 Object state transitions

The completion of a level 1 process or level 2 activity is associated with a state transition of an object. For example the completion of the level one process "Obtain listing" occurs when the property achieves a state of "fully available". Producing state transition diagrams for the main business objects of *property* and *customer* provided a useful framework for defining the key processes and as a standard analysis tool has been used widely both in workflow definition (e.g. Rusinkiewicz and Sheth, 1995) and in the service industry (Jahnke, 1995).

shows the state transitions and associated activities for the *property* (house) object.

Property object state	Event causing state change	Level 2 activity performed by negotiator
Not on market	-	-
Awaiting inspection	Initial registration	Register7 vendor
Inspected	Inspection	Inspect property
Fully available	Sale authorised by vendor	Make property available
Under Offer	Offer made by vendor	Process offer by applicant
Sold Subject to Contract	Offer accepted	Inform vendor
Sold contracts exchanged	Exchange of contracts between vendor and applicant	Confirm exchange through solicitor
Sale completion	Title and keys exchanged	Confirm completion through solicitor

Table 2: Key state transitions for the property object

5.2 Sequencing of process elements

In this particular business application the majority of process elements are serial in nature. For example, the level 2 activities of Figure 2 occur in the order given except where the listing is not obtained where the sequence is governed by a simple IF <condition> THEN <perform next activity> ELSE <perform exception activity>. This made representation of the order of activities and their enactment using database triggers straightforward.

5.3 Key areas for process improvement

Following process analysis, processes for improvement were identified for which additional facilities could be built into the system. This was mainly based on work measurement and comments from staff. For a process to warrant improvement key factors were that customer service should be significantly improved and that staff time could be reduced. Additionally improved processes should help in achieving the goals of the principal business processes. Processes selected for improvement are shown in Table 3. The importance of the WFM component of the system is indicated since many of these processes need a reminder to the negotiator to perform them to in order to achieve the correct service levels.

Process targeted for improvement	Improve data quality	Improve customer service	Reduce staff time	Reminder needed
Register customers	✓	✓		
Arrange viewings		✓	✓	✓
Perform applicant match using a map cf paper file		✓		
Perform a property match (mail-out for new property, and follow up)		✓	✓	✓
Update vendors on sale progress		✓	✓	✓
Process an offer and sale to completion	✓	✓	✓	✓
Produce property particulars		✓	✓	
Produce standard letters and forms	✓	✓	✓	✓
Improve information retrieval speed		✓	✓	
General diary and scheduling functions		✓	✓	✓

Table 3: Processes selected for improvement and the anticipated benefits

5.4 Method of workflow implementation

The method of implementing the WFMS is described briefly since this paper focuses on the process analysis methods. The WFM component of the system was developed using the same RAD tools as the remainder of the application since workflow was not the principal purpose of the application.

The WFM component consisted of a screen with a combined diary and a task list for each negotiator which contained a list of tasks to be performed on the current day. Basic information was given to the user which was necessary to complete the task, this included the type of task, the customer and/or property to which it referred, and a completion date. Tasks were completed through a conclusion button with which additional comments could be made. For each property and customer a table of significant events which had occurred for the object was displayed. This included a record of all phone or personal contact with the customer and tasks which had been completed on the task list.

The workflow scheduler was based on Oracle database triggers which were used to automatically generate tasks following events i.e. changes in state of objects within the system. Triggers (Sturner, 1995) are a standard feature of client/server Relational Database Management Systems which are efficient in that they execute on the server and do consume processing power on the client PC. When an object state changes in the database this is registered as a trigger consequent upon a SQL UPDATE statement to modify an existing field such as property price. Following a price change to a *property* object a task *Property mailout* will be produced to mail customers who were not previously given information on the property because it was not in their price range. When objects such as property or customer are created through registration and a new record created using a SQL INSERT INTO statement this will also produce tasks automatically.

6 Summary - evaluation of the initial results of the improvement programme

The system is currently undergoing evaluation in the first four branches of the test zone. Initial results indicate the following:
* good correspondence between tasks and business processes with the CSFs being successfully achieved according to branch managers
* customer and negotiator questionnaires indicate the system is effective in meeting its goals, but with further increases in efficiency of some activities such as property mailout required in future incremental releases
* capture of activities to be performed was comprehensive resulting in negotiators having to undertake more tasks than previously. Although this

increased customer service it has left an increased workload on negotiators. Ten
level 2 activities have been removed from the task list to reduce this workload
- Method of producing WFM component with Oracle triggers effective. Has
  proved easy to disable unnecessary Oracle triggers since independent for each
  task

References

Chaffey, D. (1996) Design and implementation factors in image retrieval across a Wide
Area Network - a commercial example *International Journal of Information
Management,* Vol 15, No 6.

Curtis, B., Kellner, M., Over, J. (1995) Processing modeling. *Communications of the ACM.*
Vol 35, No 9.

Davenport, T. (1993) *Process Innovation:Re-engineering work through Information
Technology.* Harvard Business School Press, Boston.

Georgakopoulos, D., Hornick,M. and Sheth, A. (1995) An Overview of Workflow
Management: From Process Modeling to Workflow Automation Infrastructure.
Distributed and Parallel databases, Vol 3 119-153.

Gruhn, V., and Wolf, S. (1995) Software Process Improvement by Business Process
Orientation. *Software Process - Improvement and Practice.* Vol 1, No 1.

Hammer, M. & Champy, J. (1993) *Re-engineering the corporation: A manifesto for
business revolution.* Harper Collins, New York.

Jahnke, B., Bachle, M. and Simoneit, M. (1995) Modelling sales processes as preparation
for ISO 9001 certification. *International Journal of Quality and Reliability
management.* Vol 12, No 9.

Karagiannis, D. (1995) Business Process Management Systems. *SIGOIS Bulletin.* Vol 16,
No. 1.

Parasuraman, A., Zeithami, V., and Berry, L. (1988) *SERVQUAL:A multiple-Item scale for
measuring consumer perceptions of Service quality. J.Retailing.* Vol 64, No 1

Rusinkiewicz, M. and Sheth, A. (1995) Specification and execution of transactional
workflows, in *Modern Database Systems: The object model, interoperability and
beyond* (ed W. Kim), ACM Press.

Scherr, A. (1992) A new approach to business processes. *In IBM Systems Journal,* Vol 32,
No 1.

Sturner, G. (1995) *Oracle 7: A User's and Developer's guide including release 7.1.*
International Thompson Computer Press, London, U.K.

Tsalgatidou, A. and Junginger, S. (1995) Modelling in the Re-engineering Process. *SIGOIS
Bulletin.* Vol 16, No 1.

Part 5 Methodological and Technical Aspects of Business Process Modelling

A Coordination-based Approach for Modelling Office Workflow

L. Yu

An Object Oriented Approach to Business Process Modelling

M. Rohloff

Business Process Reengineering with Reusable Reference Process Building Blocks

K. Lang, W. Taumann, F. Bodendorf

An Object-Oriented and Business Process-Based Meta Model of an Architecture for Management Support Systems

A. Mehler-Bicher

It's Time to Engineer Re-engineering: Investigating the Potential of Simulation Modelling for Business Process Redesign

G. M. Giaglis, R. J. Paul

A Coordination-based Approach for Modelling Office Workflow

Lei Yu

Department of Computer Science, University of Zürich, Winterthurstrasse 190, CH-8057 Zürich, Switzerland, yu@ifi.unizh.ch

Abstract. In this paper, principles and approaches derived from coordination theory to model office workflow are explored. After different definitions and models of workflow are analysed, a definition of office workflow is given. Then a goal-achieving perspective is proposed to explain office work and coordination. The goal-achieving model combines the coordination components such as goal, actor, activity and interdependence organically. So it is used as the basic paradigm for modelling office workflow. Moreover, steps for modelling workflow derived from coordination theory are discussed. The introduction of actors and positions has incorporated organisation model within workflow. Shared resources, prerequisite constraints, simultaneous constraints and reciprocal constraints are analysed and coordinated. Finally, the relations of concepts are illustrated by a conceptual architecture. And the workflow to a purchase process in a sales department is explained as an application of this approach.

Keywords. Computer supported cooperative work, coordination theory, coordination technology, office automation, workflow management systems, business process reengineering, business process modelling, groupware

1 Introduction

Many organisations struggle with the coordination of work. For example, procedures that are available on papers are not, or only partly, used in practice, work is stuck on desks of people for too long, task responsibilities are unclear and much effort is spent in corrective actions on procedural errors. To improve such situations, an understanding of the process is necessary. The business challenge is to exploit the possibilities with respect to work coordination. And workflow management is today considered as one of the key technologies for providing efficiency and effectiveness in the office. It allows the analysis of current workflow in order to detect potential bottlenecks and the design of new workflow patterns so that shortcomings can be eliminated. It is a new research area rooted in office automation, business administration, data communication, information system and computer supported cooperative work.

The central question of the research is the question of workflow analysis: how to realise the full potential of workflow in a practical situation. In order to investigate and manipulate workflow in a business, a model of the current workflow is used for documenting, understanding and communicating the coordination in business activities. The model is a natural basis for *Business Process Redesign*. The biggest change brought about by BPR is the orientation toward processes. Workflow, by ist very nature, is process oriented. This makes Workflow in general an excellent candidate for implementing the results of BPR. The relationships between Business Process Reengineering and workflow is examined in detail by Swenson and Irwin (1995).

However, attempts to build workflow tools are often delayed or even frustrated because the tool designers can find no theory upon which their tools can be built. Now analysts are confronted with a relatively new area, in which only little experience is available, let alone a well founded method. Therefore, it fits the continuous research theme of our project, STRATUM (Bauknecht and Teufel, 1993) (Teufel, Morger et al, 1995) (Seffing, 1995), to study approaches for modelling workflow by using coordination theory.

In Section 2, after different definitions and models of workflow are analysed, the definition of office workflow is given and it is concluded that to model office workflow one should consider not only the aspect of activities, but also the aspect of actors, goals, and the coordination problems between them. In Section 3, coordination theory is briefly described and a goal-achieving model is proposed to understand the relations of coordination components. And it is used as the basic paradigm for modelling office workflow. In Section 4, the steps for modelling workflow derived from coordination theory are discussed. In Section 5, a conceptual architecture of workflow is given to illustrate the relations of the concepts. In Section 6, the workflow to a purchase process in a sales department is explained as an application of this approach.

2 Definitions and Models of Workflow

With the rapid development of workflow software products, many definitions of workflow have been given in the past years. In order to promote the use of workflow through the establishment of standards for software terminology and connectivity between workflow products, a non-profit, international organisation of workflow vendors, users and analysts - the Workflow Management Coalition - was founded in August 1993. It has released a glossary (Members of Workflow Management Coalition,1994), which contains definitions of terms involved in workflow management. It divides process activities into two classes: manual process activities and workflow process activities. It defines process as a coordinated set of process activities that are connected in order to achieve a common goal. However, in the opinion of S. Joosten (1994), the glossary fails to relate the concepts to one another. He defined workflow as "a system whose elements are activities, related to one another by a trigger relation, and triggered by external events, which represents a business process starting with a commitment and ending with the termination of that commitment." He proposes that activities, roles and triggers are very important, and that the analysis and design of workflow systems should take these notions as a starting point. He has given a trigger model for workflow analysis. It makes workflow modelling different from information systems modelling, which conventionally starts with the modelling of data structures (e.g. Entity-Relationship modelling) or processes (e.g. dataflow modelling and process modelling).

Another distiction of processes is adopted by ActionWorkflow (Action Technologies Incorporation, 1993) (Medina-Mora, Winograd et al, 1992) which has divided processes into three kinds: material processes, information processes, and business processes. It follows that a business process should be viewed as a network of commitments. The commitments are contained within structures referred to as "workflow". This can be depicted as ellipse divided into four phases of interaction: Preparation, Agreement, Performance, and Acceptance as shown in Fig 2.1.

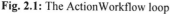

Fig. 2.1: The ActionWorkflow loop

Distributed throughout the four phases are 12 acts that define all possible work interactions. Thus within the ActionWorkflow technology, a business process is automated and managed as a network of workflow. Although ActionWorkflow is based on a speech-act theory (Winograd, 1986) (Flores, Geaves et al, 1988) (Auraaamaeki, Lehtinen et al, 1988), it keeps only the business process - the conversation between people - at the centre of attention. As Flores (1995) has pointed out, much of the ongoing research at Action today is oriented to merging these three types of processes into one seamless process that allows a workflow architecture to work easily at any relevant level of granularity. Similarly, the work of Gulla and Lindland (1994) has also relied on the distinction between workflow manifested in production process and that manifested in coordination process. They thought that ActionWorkflow lacks direct links to production activities, and that the traditional approaches such as IFO Paradigm (Input-Process-Output Paradigm) are suited to modelling the chain of production processes, although it lacks concepts for modelling coordination activities because it lacks of the involvement of actors, the exchange of information between actors, and the structuring of cooperative work process. Although both paradigms are flexible and can be extended with new concepts, it seems problematic to combine them as they are today.

Moreover, Eillis and Nutt (1993) followed the definition of workflow from S. A. Bull that a workflow is a system that helps organisations to specify, execute, monitor, and coordinate the flow of work items within a distributed office environment. They have investigated Information Control Nets (ICNs), derived from high-level Petri nets, as a tool for modelling office workflow. Later in 1994 (Ellis and Wainer, 1994), instead of choosing procedures and activities as a starting point, they choose people and goals. They were concerned with goals of people, and suggested an extended ICN model which includes the concept of goals. Other researchers (Bussler and Jablonski, 1994) (Dinkhoff, Gruhn et al, 1994) (Jablonski,1995) have proposed to integrate organisation model, activity model or data model in the workflow management environment. Yu and Mylopoulos (1994) have focused on modelling of strategic actor relationships for a richer conceptual model of business processes in their organisational setting.

It is obvious that traditional approaches are not appropriate for modelling office workflow and researchers have paid more and more attention to developing new theories and approaches on this aspect. The distinction of processes is very necessary, but these models seem still problematic to model office workflow. Office processes, unlike processes that are executed by machines, exist in social organisation settings. Organisations are made up of social actors who have goals and interests, which they pursue through a network of relationships with other actors. A richer model to office workflow should therefore include not only how activities progress, but also how the actors performing them relate to each other intentionally, i.e., in terms of concepts such as goal, dependency and coordination. Our work will study a new approach for modelling office workflow which is based on coordination theory and can overcome some of the shortcomings of traditional approaches. First of all, we define office workflow as follows:

Definition 1 (Workflow) A workflow is a coordinated set of interdependent activities which are performed by actors in an organisation in order to achieve a set of common goals.

3 Coordination Theory and A Goal-achieving Perspective

There are many coordination problems in an organisation such as shared resources, producer/ consumer relationship, simultaneous constraints, task/subtask problems. In the Eighties there was an increasing interest in studying how the activities in complex systems could be coordinated. This was a new area of research, though it was based on several other research areas such as computer science, organisation theory, management science, economics, linguistics and psychology. In 1988 Malone finished a working paper "What is Coordination Theory ?" and he became one of the most important pioneers in this new discipline.

Early Malone and Crowstone (1990) have defined coordination "broadly" as "the act of working together harmoniously", and the basis of their work is to find a body of principles about how activities can be coordinated, that is, how actors can work together. And they summarised the components of coordination as shown in Table 3.1.

Components of coordination	Associated coordination process
Goals	Identifying goals
Activities	Mapping goals to activities
Actors	Assigning activities to actors
Interdependences	"Manage" interdependences

Table 3.1: Components of coordination

Goals are the reasons of the actions. *Actors* are the subjects who are responsible for performing the actions. An actor is any person, machine or group that is capable of performing activities or being responsible for activities. *Activities* are the performed actions. *Interdependences* will be considered among correlated activities in a cooperative working environment.

Using the components of coordination, Malone and Crowstone (1990, 1994) gave a narrow definition of *coordination* as "the act of managing interdependences between activities performed to achieve a goal".

Traditional modelling approaches pay more attention to activities analysis, and less on the involvement of actors. In our opinion, much of the office work in organisations is goal-based rather than activity-based, and people do whatever is necessary to attain these goals. So we shall focus not only on activities, but also goals, actors and the interaction between them in modelling office workflow.

In order to understand the the relationships of the components of coordination we shall propose a goal-achieving perspective. This perspective starts with selecting an overall goal. This goal is divided into activities, which demand actors and other resources. The actor performs the activity, the activity produces a result. The result ought to be compared with the goal. Then a decision m will be made by the actor as to what to do next and how to do it. The process is modelled in Fig 3.1.

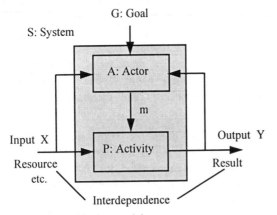

Fig. 3.1: Goal-achieving model

This model represents a basic working unit in an office which we call a goal-achieving model. From a view of system theory (Mesarovic and Takahara, 1989), we can call it a goal-achieving system. A system is recognised as a goal-achieving system if for any given input the corresponding output is determined in reference to some internal activities of the system aimed at attaining a goal. In an organisation, the basic work items called activities, must be performed by people (actors) in a certain sequence to attain the goals. So we can regard an organisation as a complex system consisting of many goal-achieving systems that share a set of common goals. In order to attain the overall goal of the organisation, actors must work together and coordinate with each other. Let S_i represent sub-system i, X_i represent input, Y_i represent output and G_i represent goal of sub-system i. Then interaction of the sub-systems in an organisation is modelled using this notation in Fig 3.2. Because the output/result of one sub-system may be an input/resource of another sub-system, we can call this interdependence among sub-systems. This is a general description of a goal-achieving perspective and it corresponds to Malone and Crowstone's description in Table 3.1. So it is very useful in understanding coordination theory.

The task for workflow is to describe the coordination and performance of work undertaken in an organisation. Now that the goal-achieving model not only covers the four components of coordination, but also combines them organically, we will use it as a fundamental paradigm in modelling office workflow.

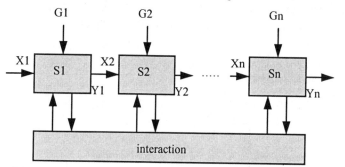

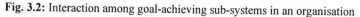

Fig. 3.2: Interaction among goal-achieving sub-systems in an organisation

4 Modelling Office Workflow by Using Coordination Theory

Step 1 Goal identification and position determination

Among the four components, the interdependences, actors, and activities will evolve continually, but goals remain more stable. Therefore, it is reasonable to begin our work with identification of goals of an organisation.

There are many methods used to refine goals and decompose them. Anton, McCracken et al (1994) have presented experiences in applying the goal decomposition and scenario analysis model in the context of business process reengineering. For examples, these methods can be scenarios analysis, bottom-up method, top-down method, hybrid of bottom-up and top-down method.

Instead of naming a person directly as the actor for achieving a goal, we can specify that a goal is to be achieved by a particular position. The process of determining which goals should be achieved by one position is termed position determination. And we define position as follows:

Definition 2 (Position) A position is a set of related abilities/competence and responsibility which are necessary for achieving a goal or completing an activity.

For example, in a sales department, the overall goals are to expedite sales and maximise customers' satisfaction, and they can be divided into sub-goals such as to maximise profit, to maximise customers' satisfaction, to minimise risk in trade etc.. Then the positions such as salesman, warehouse worker and accountant can be determined. Each position is responsible for the given goals.

242

A position may be associated with a group of actors rather than a single actor. Also, one actor may have many positions within an organisation. For example, there are ten persons who fill the position of salesman, and Jim (one person) fills the positions of both salesman and credit staff member. In workflow, goal will be graphically represented by a circle and position by a rectangle. And the goal-achieving process is simply modelled by drawing a arc between goal and position as shown in Fig 4.1.

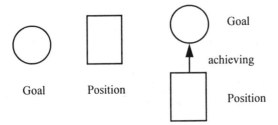

Fig. 4.1: Legends of goal and position

Step 2 Mapping goals into activities and actor selection

In this step, the problem is how a position meets his goals; i.e., how to find out which activities are performed under the responsibility of a position to achieve his goals. There are two kinds of activities. One is called complex activity, which has sub-activities. Another is elementary activity, which can not be broken down any longer. The methods used in goal decomposition are suited to activity decomposition. A complex activity can be broken into its components, and components are broken down into sub-components until they become elementary activities. Sometimes one position can not complete the complex activity. In this case the decomposed activities can be grouped into new positions. One criterion for grouping activities into position is to minimise the difficulties of managing the interdependences. For example, activities with the strongest interdependences are often grouped into the same position so that the positions have weaker interdependences.

To complete the activities, actors will be selected to fill the positions. This process is called *actor selection*. For example, the manager will select who is best qualified for the position of a salesman who needs a special ability to negotiate with customers. An authority-based decision-making process might be used in which the manager simply assigns activities to people who agree to accept such assignments. But actors play very important role in achieving goals so that actor selection is always taken into serious consideration in an organisation. Particularly when many candidates compete for an important position which has multiple criterions, a group decision-making process might be adopted in which alternatives are proposed, evaluated and chosen by a group of decision makers. Cheng and Yu (1992) have proposed an interactive approach to solve these kind of problems. Since an actor is a kind of resource, actor selection is also a kind of "resource allocation activity".

Step 3 Interdependences analysis

If there is no interdependence, there is nothing to coordinate. Malone and Crowstone (1990) have proposed a preliminary list of types of interdependences and coordination processes that can be used to manage them. These interdependences are shared resources, prerequisite constraints and simultaneous constraints. In addition to the above types, we will introduce a new kind of interdependence called reciprocal constraints in modelling workflow. Then four kinds of interdependences will be further discussed below.

Shared resource

Definition 3 (Resource) Resource is something passively utilised during activity performance.

Whenever multiple activities share some limited resource (e.g., money, tools, data or an actor's time), then a resource allocation process is needed to manage the interdependences amongst these activities. Resource allocation is perhaps the most widely studied of all coordination processes. Control of resources is intimately connected with personal and organisational power: those who control resource have power and vice versa. In general, organisational theorists emphasise hierarchical resource allocation methods where managers at each level decide how the resource they control will be allocated among the people who report to them. Examples of coordination process for allocating resource are "First come/first served", priority order, budgets, managerial decision, market-like bidding. What coordination process will be chosen is decided by the actor, guided by his goals.

Resource, represented by ellipse in workflow, are repositories of the data, people or material produced or consumed by the activities. An incoming flow represents an update or produce to resource, for example, data is a kind of resource, and an incoming flow means either in the form of additional data, a modification to existing data, or the deletion of data. An outgoing flow represents access to or consumption of the resource. These are shown in Fig 4.2.

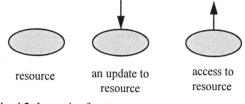

resource   an update to   access to
                  resource   resource

Fig. 4.2: Legends of resource

Prerequisite constraints

In workflow, some activities depend on the completion of others before they can start. This kind of interdependence is termed ***prerequisite constraints***. It deals with starting conditions of an activity.

Definition 4 (Event) An event is something that happens, something that occurs (Joosten, 1994).

Definition 5 (Object) An object is something that is (or is capable of being) seen, tourched, or otherwise sensed (Joosten, 1994).

An event is carried by an object. It is produced by one activity and consumed by other activities. So it is the basic component to model prerequisite constraints. There are two kinds of prerequisite constraints. We also term them And-join and Or-join (Members of Workflow Management Coalition, 1994) in office workflow.

When two or more parallel events are all needed for triggering an activity, they converge into a single common flow of control. This prerequisite constraint is an *And-join*. When any of two or more events causes the beginning of an activity , this prerequisite constraint is an *Or-join*. In the case of Or-join there is no synchronisation of events from each of the two or more activities to the single activity. And-join and Or-join are graphically represented in Fig 4.3.

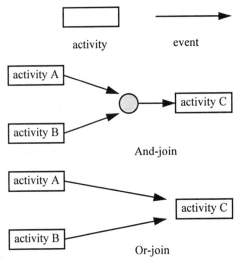

Fig. 4.3: Legends of prerequisite constraints

Simultaneous constraints

Another common kind of interdependence between activities is that they need to occur at the same time (or can not occur at the same time). This is termed *simultaneous constraints*. Synchronisation analysis is also suited to modelling them.

When a single event splits into two or more at the same time in order to perform activities in parallel, this simultaneous constraint is an *And-split*. When confronted by multiple branches of events to activities, an actor makes a decision upon which event occurs. This simultaneous constraint is an *Or-split*. These two kinds of simultaneous constraints are graphically represented in Fig 4.4.

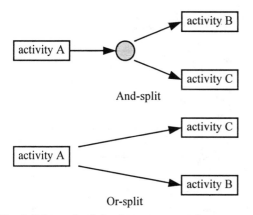

And-split

Or-split

Fig. 4.4: Legends of simultaneous constraints

Reciprocal constraints

In some cases, where each activity depends on the other, or where each activity requires inputs from the other, this is called *reciprocal constraints*. For example, in negotiation, both sides depend on each other before a set of common conditions can be reached. This kind of interdependence is modelled by a two-way flow, or two flows in opposite directions.

5 A Conceptual Architecture of Workflow

In this section, we present the conceptual architecture of workflow using the Entity-Relationship model to illustrate the definitions. The architecture shown in Fig 8 builds upon the general concepts introduced in the previous sections. It lays out the workflow conceptual entities and their relationships.

The Entity-Relationship model is a high level semantic model using nodes and arcs; this model has proved to be useful as an understandable specification model and has been implemented within E-R database. A labelled rectangle denotes an entity set; a labelled arc connecting rectangles denotes a relationship between the corresponding entity sets. In Fig 5.1, the box labelled "actor" denotes an entity set of actors that may actually be a table of actor names and their attributes. Likewise, the box labelled "position" denotes an entity set of positions and may be a table of position names and their attributes. There is an arc connecting these two boxes because there is relationship called "fill" between these two entity sets. This arc is labelled with the relationship name, and a denotation of "M" and "N" indicating that this is a many to many relationship. Therefore, an actor can fill many positions, and a position can be filled by many actors.

246

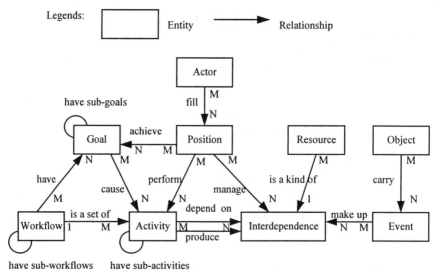

Fig. 5.1: A Conceptual Architecture of Workflow

6 An Example

To explain the application of the coordination-based approach, we will model the workflow to a purchase process in a sales department by using the steps and the legends discussed in previous sections.

At first the overall goal of the sales department is divided into sub-goals and positions such as salesman, warehouse worker and accountant are determined. Then people are selected to fill these positions. Their name can either be written down near the positions' name (see in Fig 6.1 the position of salesman) or in another corresponding table. In order to minimise risk in trade, a salesman must check the customer's credit. So the goal to minimise risk is mapped into credit-checking activities. To check the credit, the salesman needs to know credit information about customers. So resource (customers' information) is consumed by this activity. If the salesman is satisfied with the checked result, then he will decide to sell the products to the customer and allocate limited products among the customers in order to maximise profit and customers' satisfaction. Of course, he will negotiate with the customer. This process of negotiation is a reciprocal constraint. If they reach a agreement on the purchase after negotiation, then the salesman will instruct a warehouse worker to ship products and the accountant to record the payment. This is an And-split constraint. The customer will pay for the products if and only if he receives both the products and the bill from the warehouse worker and accountant respectively. This is an And-join constraint. The arcs connecting goals and positions represent the goal-achieving process. The events between positions make actors in communication visible. Goals, positions, actors, activities, resource and constraints are all graphically represented in Fig 6.1.

Legends are shown in Fig 4.1, Fig 4.2, Fig 4.3 and Fig 4.4.

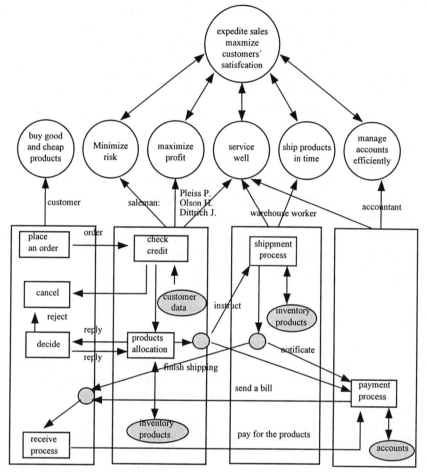

Fig. 6.1: Workflow to a purchase process in a sales department

7 Conclusion

In this paper, principles and approaches that model office workflow by using coordination theory are explored. After different definitions and models of workflow are analysed, the definition of office workflow is given and we conclude that to model office workflow one should consider not only the aspect of activities, but also the aspect of actors, goals, and the coordination problems between them. We regard office work as a goal-achieving activity and organisation as a complex system consisting of many goal-achieving sub-systems. Coordination is the act of managing the interdependences between these activities. The goal-achieving model covers all of the four coordination components, i.e., goal, actor, activity and interdependence, and combines them organically. So it is used as the basic paradigm for modelling office workflow. The goal-achieving perspective corresponds to the coordination theory. Therefore, the coordination-based approach has a theoretic basis. The introduction of actors and positions has incorporated organisation model within workflow. Shared resources, prerequisite constraints, simultaneous constraints and reciprocal constraints can be coordinated and modelled. And the communication of actors is also made visible in this model. Therefore, it helps to overcome some of the shortcomings of other modelling approaches.

Workflow is a fertile area for research and development. There are significant problems to be solved to ensure its future success. There is a need for implementing a prototype of the coordination-based approach. Then the issues such as workflow analysis, simulation and redesign can be studied. Moreover, this approach can be evaluated and improved in applications.

Acknowledgements. The author wish to thank Prof. Dr. Kurt Bauknecht, Mr. Jan Seffinga, Mr. Othmar Morger, Mr. Andrew Hutchison and Dr. Stephanie Teufel for their discussion and help in writing this paper.

References

Action Technologies Inc. (1993). *ActionWorkflow Analyst User's Guide (Version 2.0)*, .

Anton, A. I., McCracken, W. M., and Potts, C. (1994). "Goal Decomposition and Scenario Analysis in Business Process Reengineering." *Proceedings of 6th International Conference on Advanced Information Systems Engineering, CAiSE'94*, Utrecht,The Netherlands, 94-104.

Auramaeki, E., Lehtinen, E., and Lyytinen, K. (1988). "A Speech-Act-Based Office Modelling Approach." *ACM Transaction on Office Information Systems*, 6(2), 126-152.

Bauknecht, K., and Teufel, S. (1993). "STRATUM: Strategische Unternehmensführung - Methoden und Konzepte zur Unterstützung der kooperativen Aufgaben des Informationsmanagements." *belligtes Forschungsgesuch (Schweizerischer Nationalfonds), Nr. 1214-37458.93*, Institut für Informatik der Universität Zürich,.

Bussler, C., and Jablonski, S. (1994). "An Approach to Integrate Workflow Modeling and Organization Modeling in an Enterprise." *Proceedings of third Workshop on Enabling Technologies: Infrastructure for Collaborative Enterprises*, Morgantown,West Virginia, 81-95.

Cheng, M. X., and Yu, L. (1992). "Group Evaluation with Multiple Objectives and Its ICM Approach." *AMSE Review*, 20(1), 45-52.

Dinkhoff, G., Gruhn, V., Saalmann, A., and Zielonka, M. (1994). "Business Process Modelling in the Workflow Management Enviroment Leu." *Proceedings of Conference on Entity-Relationship Approach, ER'94*, 46-63.

Ellis, C. A., and Nutt, G. J. (1993). "Modelling and Enactment of Workflow Systems." *14th International Conference on Application and Theory of Petri Nets 1993*, Chicago,Illinois,USA, 1-16.

Ellis, C. A., and Wainer, J. (1994). "Goal-based models of collaboration." *Collaborative Computing*, 1(1), 61-86.

Flores, F., Graves, M., Hartfield, B., and Winograd, T. (1988). "Computer Systems and the design of Organizational Interaction." *ACM Transactions on Office Information Systems*, 6(2), 153-172.

Flores, R. F. (1995). "The Value of a Methodology for Workflow." , http://www.actiontech.com/market/papers/method5.html, WWW, 1-3.

Gulla, J. A., and Lindland, O. I. (1994). "Modelling Cooperative Work for Workflow Management." *6th International Conference,Advanced Information Systems Engineering,CAiSE'94*, Utrecht,The Netherlands, 53-65.

Jablonski, S. (1995). "Workflow-Management-Systeme: Motivation, Modellierung, Architektur." *Informatik Spektrum*, 18(1), 13-24.

Joosten, S. (1994). "Trigger Modelling for Workflow Analysis." *Proceedings of Ninth Austrian-Informatics Conference, Con'94 : Workflow Manangement - Challenges,Paradigms and Products*, Linz,Austria, 236-247.

Joosten, S., and Brinkkemper, S. (1995). "Fundamental Concepts for Workflow Automation in Practice." *Proceedings of Conference ICIS'95*, AMSTERDAM.

Malone, T. W. (1988). "What is Coordination Theory ?" *CISR WP NO.182/Sloan WP No.2051-88*, Center for Information Systems Research, Sloan School of Management, Massachusetts Institute of Technology.

Malone, T. W., and Crowstone, K. (1990). "What is Coordination Theory and How Can It Help Design Cooperative Work Systems?" *Proceedings of the Conference on Computer-Supported Cooperative Work,CSCW'90*, Los Angeles,CA, 357-370.

Malone, T., and Crowstone, K. (1994). "The Interdisciplinary Study of Coordination." *ACM Computing Surveys*, 26(1), 87-119.

Medina-Mora, R. (1992). "ActionWorkflow Technology and Application for Groupware." *Proceedings of Conference on GroupWare, GroupWare'92*.

Medina-Mora, R., Winograd, T., Flores, R., and Flores, F. (1992). "The Action Workflow Approach to Workflow Management Technology." *Proceedings of the Conference on Computer Supported Cooperative Work, CSCW'92*, 281-288.

Members of W. M. C. (1994). "Glossary - A Workflow Management Coalition Specification." , Workflow Management Coalition, Brussels,Belgium.

Mesarovic, M. D., and Takahara, Y. (1989). *Abstract Systems Theory*, Springer-Verlag.

Seffinga, J. (1995). "Entwicklung eines theoriebasierten Geschaeftsprozessmodells", Semesterarbeit im Fach Wirtschaftsinformatik, Institut für Informatik der Universität Zürich.

Swenson, K. D., and Irwin, K. (1995). "Workflow Technology: Tradeoffs for Business Process Re-engineering." *Proceeding of Conference on Organizational Computing Systems*, Milpitas, California, U.S.A., 22-29.

Teufel, S., Morger, O., and Seffinga, J. (1995). "Strategische Unternehmensführung - Methoden und Konzepte zur Unterstützung der kooperativen Aufgaben des Informationsmanagements, STRATUM, Institut für Informatik der Universität Zürich.

Winograd, T. (1986). "A Language/Action Perspective on the Design of Cooperative Work." *Reading*, 23, 623-652.

Yu, E. S. K., and Mylopoulos, J. (1994). "From E-R to ´A-R´ - Modelling Strategic Actor Relationships for Business Process Reengineering." *Proceedings of Conference on Entity-Relationship Approaches, ER'94*, 548-565.

An Object Oriented Approach to Business Process Modelling

Michael Rohloff

Danet GmbH, Hansastraße 32, 80686 München, michael.rohloff@danet.de

Abstract. This paper introduces an object oriented approach to business process modelling. The approach integrates decisions on the organisational design with the information systems development. Based on the understanding of business processes as a customer - supplier relationship a general process model is introduced which summarises fundamental characteristics of business processes. These characteristics are the ground for an object oriented approach which extends the oo modelling-technique OMT with features for business process modelling. An example illustrates the principal approach, the main modelling steps and the methodical support. It is based on an outside (macro) and an inside (micro) view on business processes.

1 Business Process Modelling and Object Orientation

1.1 A General Business Process Model

Within the last years a still growing importance in organisational design is given to business processes and the flow of work (Davenport 1993, Davenport/Short 1990, Hammer/Champy 1993, Krickel 1994 etc., for an overview on tools see Hess/Brecht 1995). A wide range in scope and content of a business process can be observed in industrial practise and academic theory. The span is from a strategic -oriented perspective of a process which contains the whole value chain of a business to a merely operational perspective with only the activity of one person or system in the focus (e.g. Ferstl/Sinz 1993, Müller-Luschnat et. al. 1993, Scheer 1994 etc.). This is due to the fact that a variety of persons, like managers, users, systems analysts, and developers are involved in organisational design and information systems development, each with different tasks and a distinct view on business processes.

Despite this differing view on processes, a core understanding of characteristic features and basic principles of a process can be derived which are comprised in a general model of business processes (figure 1). A process is a flow of activities to produce an output of value to the customer (Hammer/Champy 1993). Each process can be viewed as an exchange of services embedded in a customer-supplier relationship (see also Medina-Mora et al. 1992, Scherr 1993). The customer requests the desired output whereas the supplier performs the process. The process activities are the means to transform process input (material, information) into the desired output. During its performance the process changes states which are triggered by events. The process flow is determined by business rules (conditions).

Although a process can be viewed on different levels of abstraction, depending on the domain of the analysing person, each process follows these outlined principles.

To view each process as a customer-supplier relationship is the fundamental principle of the approach described in this paper: it ensures a strong orientation on customer requirements and a clear assignment of the responsibility for the process. According to the transaction cost approach (Williamson 1975, Picot 1982, for transfer to business process modelling see Medina-Mora et al. 1992, Scherr 1993) each process can be structured in the phases opening, negotiation, performance, and acceptance (customer satisfaction).

To summarise, the main characteristics of a process can be described as a(n):
a) exchange of services between a customer and a supplier,
b) input-output transformation to produce business value,
c) logic flow of activities which have to be co-ordinated,
d) co-operative performance by a number of actors,
e) use of resources for the performance.

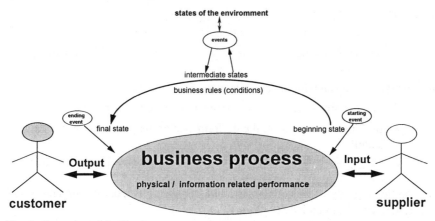

Fig. 1: General model of business processes

1.2 Basic Modelling Principles (Object Oriented Meta Model)

The structure as well as the behaviour (dynamics) of a business process can be efficiently modelled with an object oriented approach. As already pointed out by the general business process model, the core description of a process and its flow is based on three main entities:
- the process itself which can be subdivided in process steps,
- events which trigger processes,
- states which show the actual status and the change along the process flow.

These entities can be very well addressed with object orientation. Putting the process into the context of the business and the organisation, further entities can be derived.

The object oriented meta model in figure 2 shows the object classes relevant for modelling business processes (for the OMT notation refer to Rumbaugh et al. 1991). The object classes can be allocated to three groups, which address:

- ◆ business objectives, tasks, and performance measurements:
  - business objective: gives the frame and target system for the processes
  - task: describes the tasks/functions realized by processes
  - reference value: performance criteria for the measurement of processes
  - performance: result or output of a process
- ◆ process input and resources allocated to the process:
  - process input: input of a process are information, tools, and material
  - resource: resources necessary to perform a process like personnel and systems
- ◆ personnel involved with the performance of the process:
  - process owner: responsible for the process and its performance
  - performer: involved in the performance of the process as a customer or supplier
  - role: a set of functions assigned to a performer in a specific process

An additional group of object classes addresses the corresponding organisational structure which is not included in figure 2.

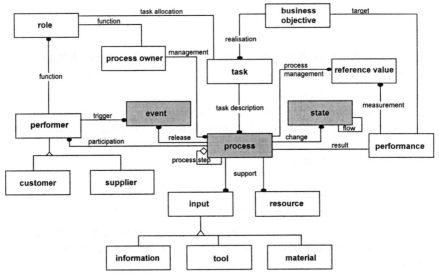

Fig. 2: OMT meta model for business process modelling (process flow perspective)

The meta model is the basis for the description of business processes and their performance within a business organisation. It is used as a cross reference for the entities related to a business process. Primary focus is on the core entities process/process step, event, and state in order to describe the business flow using:
- object models to describe object classes and their relationships,
- interaction diagrams to address the exchange of events among object classes,

- and for the dynamic modelling state charts to describe the behaviour of object classes in terms of events, states and the logic of flow.

The OGM approach outlined here (OGM is the German abbreviation of object oriented business process modelling) is based on OMT (Rumbaugh et al. 1991) as a widely used object oriented modelling technique. OMT and other popular oo-methods (for a comparison see Stein 1993) focus on the domain analysis and the information systems design (except for the use case driven approach of Jacobson 1992, 1994 and the SOM method of Ferstl/Sinz 1995 which also address business process modelling). Presently, these methods are still weak in terms of modelling the overall process flow since the focus is on the behaviour of single objects which usually cover only a small part of the entire process. Thus, the explanation of the interaction of objects in the context of a business process is limited. For this reason the OGM approach enhances OMT with further techniques to adequately address business process modelling.

The OGM approach consists of two modelling steps, the macro and the micro view of the process. The "order processing" of a customer to order manufacturer is used as an example to demonstrate the procedure of the OGM approach.

2 Macro View of the Process

The macro view takes a global process perspective from an outside point of view. At this stage all features of a process are analysed with the focus on its interfaces to the environment. Each process is modelled as a unique macro object which describes features and requirements of the process in an overview level. However, first insights into the process can be gained which are the ground for a detailed analysis in the micro view of the process.

Starting point are the business objectives and the alignment of the supporting processes. This business - objective driven description of processes is the fundamental principle at all stages of the modelling process. Main activities are:
- identification of the process and of the customer-supplier relationship,
- description of the process structure (sub-processes) and the input-output performance,
- determination of the persons which participate in the performance of the process.

The proceeding results in a task-objective frame, an interface model, and a record of the customer-supplier interaction.

2.1 Task Objective Frame

Basis for the identification of processes and the derivation of essential features are the general tasks and objectives (desired result/performance of a process in combination with quality measurements) for the specific business domain under analysis. They both build the framework for the design of business processes. In our example the objective of the make to order manufacturer to sell products is realised by an order management and can be further qualified by targets like satisfied customer or order processing time less than 3 weeks. A goal oriented reference is fundamental for an efficient process design and is an essential part of the OGM approach. Consequently, a line up of the process design with the business objectives is performed at each level of the process modelling and described in a business objective - action tableau. Figure 3 shows an extract of a possible set of actions for the example "order management".

Objective	task	target	action
• selling of make to order products	• order management	• satisfied customer • order - processing time less than 3 weeks	• intensive customer contact • integration of production planning tasks

Fig. 3: Business objective - action tableau (extract)

2.2 Process Interface

The process interface model describes the business process and its interactions with the environment. It shows the interface to other objects of the business domain. The business process is modelled as an object class itself. This process class is a conceptual class, viewing a process as a macro object which comprises the interaction of several objects which make up the overall performance of the process. Customer and supplier are also modelled as objects and set into relation to the process. This results in an interface model, similar to the object model of OMT (Rumbaugh et. al 1991, p. 21 f.) except that processes are shown as an oval in the diagram in order to distinguish them from other objects and to stress that each process object can be further decomposed (which is demonstrated in part 3.4).

Fig. 4: Object model of the business process "order processing" (interface model)

Although this interface model in itself is very simple, it gives the frame for a detailed analysis of the process. In addition, put into context the combination of all interface models for the main processes of a given business domain describes the overall market performance and the business partners.

2.3 Customer Supplier Interaction

A first insight on the process flow can be obtained with the analysis of the communication between customer and supplier (figure 5). All events which are exchanged between the business partners during the performance of the process are collected and are put into a time order according to the structure of the process into the phases opening (o), negotiation (n), performance (p), and acceptance (a).

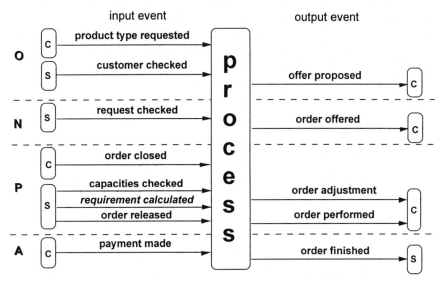

Fig. 5: Customer supplier interaction

Events are differentiated from a process point of view into input and output events. The result of this analysis is described in a customer supplier interaction diagram, like the one for the example "order processing" given here (c = customer, s = supplier, arrows indicate the directions of events from the process point of view). It is supplemented by an event list (not included here) which further specifies each event.

258

3 Micro View of the Process

Based on the process requirements resulting from the macro process perspective the second modelling step analyses the structure and the flow of the process in more detail. The comprehensive view of the customer-supplier relationship is decomposed into sub-processes and differentiated according to the performers of the process.

3.1 Performer Interaction and Process:

Each process can be split into sub processes which make up the performance and processes which have only supporting co-ordination functions. For further analysis of a process it proved to be helpful to identify all persons (performers) which are involved in the performance of the process under question. Each performer plays one or more specific roles in the process. Ordering all performers according to their participation in the process results in a customer-supplier chain. For the process "order processing" a possible performer chain is:

customer

  sales officer

    construction engineer

    disponent

      stock manager

      purchaser

        materials supplier

    production manager

      foreman

        worker

Each interaction of two performers in the value chain can be viewed as a distinct customer-supplier relationship (each indentation marks such a relationship; performers from outside the analysis domain are in italics). Thus, additional customer-supplier interactions can be derived and described in separate interaction diagrams for each bilingual communication along the customer-supplier chain.

A typical design decision is the allocation of roles to specific types of persons, which are authorised to perform the tasks related to these roles assigned. For our example, a means for reducing lead times of the "order processing" (refer to business objective - action tableau, fig. 3) could be the decision for an integration of tasks, e.g. to assign the role of "disponent" and "foreman" to one performer "production planner". A performer role tableau documents which roles are allocated to a performer including the competence assigned (figure 6).

performer	role	competence
production planner	disponent	materials planning quantity of supply.......
	foreman	....

Fig. 6: Performer role tableau (extract)

3.2 Process Flow (Dynamic Model of the Process Steps):

In addition to the identification of the customer-supplier chain the process can be divided in easy to grasp sub-processes (s. Gaitanides 1983, p. 75 f.). Criteria for sub-processes are (s. Wittlage 1993, p. 207):
- change of the performer,
- change of the object of performance,
- split up or union of working objects,
- or the change of working tools/documents.

The process "materials planning" is an example for a sub-process from the overall process "order processing". Each sub-process can be represented in a process flow diagram, as it is illustrated for the "materials planning" in figure 7.

260

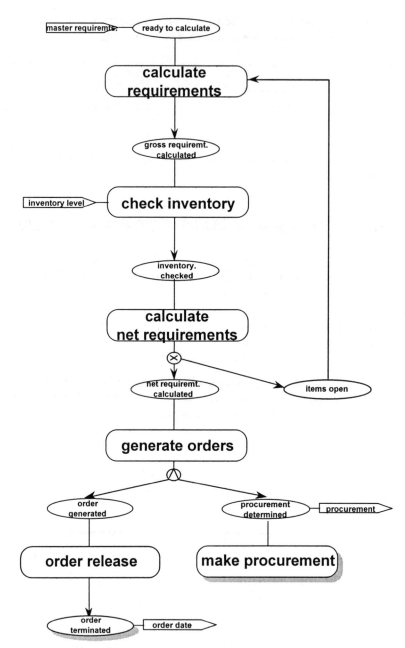

Fig. 7: Process flow diagram "materials planning"

The diagram shows the process steps (rounded rectangle), their beginning and final states (ellipse form) and the corresponding events (arrows). The arrow lines show the flow of control and the circle indicates a split up (x = exclusive or, V = or, A = and). Business rules help to identify the conditions on which a transition between states takes place (e.g. structuring in event condition action (ECA), s. Knolmayer/Herbst 1993, Herbst/Knolmayer 1995). A shadowed element denotes a link to another process flow diagram.

By means of drawing the process flow diagram an overall view on the process can be gained and a first requirements definition for the process flow can be build up, which will be detailed in the interaction of business objects in the following modelling steps (s. section 3.4).

3.3 Process Interaction

Since the overall process "order processing" is composed of several sub-processes (see section 3.2) like "materials planning" and „materials logistics", process interaction diagrams are used to visualise the exchange of events between sub processes. Each sub-process is listed with its process steps as defined in the process flow diagram (compare figure 7 "materials planning"). Each arrow indicates one event flow between process steps of different processes.

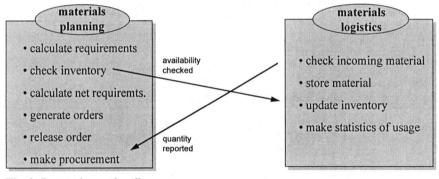

Fig. 8: Process interaction diagram

3.4 Object Model of the Process

The interface model supplemented by the process flow diagram is the ground for the identification and modelling of business objects, which generate the process flow described in the process flow diagram. According to the fundamental characteristics of a process (s. section 1.1) several types of business objects, each type with distinct and characteristic features, can be distinguished.

Each process is an assembly of several types of business objects listed in figure 10. By this means a process, so far modelled as a macro object, can be decomposed into the co-operating business objects. Each type of a business object has distinct properties which characterise the type. For example:

- *activity objects* usually feature relatively complex operations,
- *information objects* have only little functionality assigned to the operations and are mainly characterised by their attributes,
- whereas *co-ordination objects* are often involved in the performance of several processes.

An object model of the business objects for the example "materials planning" is modelled in OMT in figure 9.

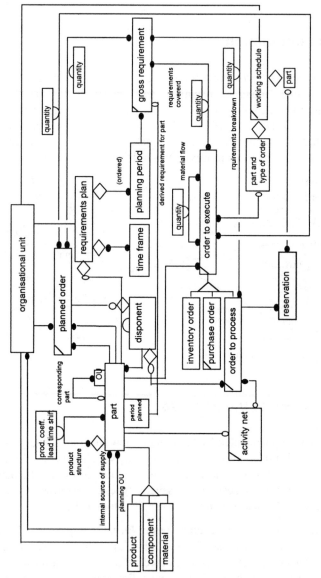

Fig. 9: Business objects of the sub process „materials planning"

process characteristic	type of business objects	example "order processing"
a) exchange of services	performer	disponent
b) input output transformation	input object output object	not in example part
c) logic flow of activities	activity object information object	requirements plan, gross requirement planning period, reservation, time frame
d) collaboration of actors	co-ordination objects	planned order, order to execute
e) use of resources	resource object	working schedule, activity net

Fig. 10: Type of business objects

The classification of object types provides a general pattern how to model business processes. At first with the model of the business objects we are at the same level of abstraction as common oo-methods. Figure 9 illustrates this for the process "materials planning". In addition, the identified business objects are modelled in state charts and event flow diagrams (straightforward OMT for reasons of place not included here, details in Rohloff 1996, p. 108 f.). Object models and dynamic models of each business object in combination with the process flow diagram describe the performance of a process and build the requirements definition for the applications development.

4 Conclusions

This paper described the OGM approach for business process modelling in extracts (more detailed in Rohloff 1996). It was demonstrated how object oriented modelling techniques can be extended in order to address business process design. By means of switching from a macro to a micro view of a process an outside to inside procedure or vice versa can be applied which makes it possible to model process requirements on different levels of abstraction with focus on management, users or systems analysts/developers. In order to visualise process modelling, additional techniques which are appropriate for these perspectives (like value chain representation for management) are added which supplement the procedure explained in this paper.

References

Davenport, T./ Short, J. (1990), The New Industrial Engineering: Information Technology and Business Process Redesign, Sloan Management Review (1990), p. 11-27

Davenport, T. (1993), Process Innovation: Reengineering Work through Information Technology, Boston

Ferstl, O./ Sinz, E. (1993), Geschäftsprozeßmodellierung,in: Wirtschaftsinformatik 35(1993) 6, p. 589-592

Ferstl, O./ Sinz, E. (1995), Der Ansatz des Semantischen Objektmodells (SOM) zur Modellierung von Geschäftsprozessen, in: Wirtschaftsinformatik 37(1995)3, p. 209-220

Gaitanides, M. (1983), Prozeßorganisation: Entwicklung, Ansätze und Programme prozeßorientierter Organisationsgestaltung, München

Hammer, M./ Champy, J. (1993), Reengineering the Corporation, New York

Hess, Th./ Brecht, L. (1995), State of the Art des Business Process Redesign, Wiesbaden

Herbst, H./ Knolmayer, G. (1995), Ansätze zur Klassifikation von Geschäftsregeln, in: Wirtschaftsinformatik 37(1993)2, p. 149-159

Jacobson, I./ Christerson, M./ Jonsson, P./ Övergaard, G. (1992), Object-Oriented Software Engineering: A Use Case Driven Approach, Wokingham et al.

Jacobson, I./ Ericsson, M./ Jacobson, A. (1994), The Object Advantage: Business Process Reengineering with Object Technology, Wokingham et al.

Knolmayer, G./ Herbst, H. (1993), Business Rules, in: Wirtschaftsinformatik 35(1993)4, p. 386-390

Krickl, O.C. Ed. (1994), Geschäftsprozeßmanagement, Heidelberg et al.

Medina-Mora, R./ Winograd, T./ Flores, R./ Flores, T. (1992), The action workflow approach to workflow management technology, in: ACM (Ed.): CSCW Proceedings, Toronto 1992, p. 1-10

Müller-Luschnat, G./ Hesse, W./ Heydenreich, N. (1993), Objektorientierte Analyse und Geschäftsvorfallmodellierung, in: Mayr, H. (Hrsg.): Objektorientierte Methoden für Informationssysteme, Berlin et al.

Picot, A. (1982), Transaktionskostenansatz in der Organisationstheorie: Stand der Diskussion und Aussagewert, in: DBW 42(1982)2, p. 267-284

Rohloff, M. (1996), Objektorientierte Modellierung betrieblicher Abläufe im OGM Ansatz, in: EMISA-Forum, Mitteilungen der GI-Fachgruppe „Entwicklungsmethoden für Informationssysteme und deren Anwendung", (1996)1, p. 100-110

Rumbaugh, J./ Blaha, M./ Premerlani, W./ Eddy, F./ Lorensen, W. (1991), Object oriented Modelling and Design, Englewood Cliffs, New Jersey u. a.

Scheer, A.-W. (1994), Wirtschaftsinformatik: Referenzmodelle für industrielle Geschäftsprozesse, Berlin u.a.

Scherr, A. (1993), An new Approach to Business Processes, in: IBM Systems Journal, 32 (1993)1, S. 80-98

Stein, W. (1993), Objektorientierte Analysemethoden - ein Vergleich, in: Informatik-Spektrum 16(1993)6, S. 317-332

Williamson, O. E. (1975), Markets and Hierachies: Analysis and Antitrust Implications: A Study in the Economics of Internal Organisation, London

Wittlage, H. (1993), Unternehmensorganisation: Einführung und Grundlegung mit Fallstudien, Herne

Business Process Reengineering with Reusable Reference Process Building Blocks

Klaus Lang, Wolfgang Taumann and Freimut Bodendorf

University of Erlangen-Nuremberg, Department of Information Systems, Lange Gasse 20, 90403 Nuremberg, Germany, lang@wi2.wiso.uni-erlangen.de, rz94-118@wsrz1.wiso.uni-erlangen.de, bodendorf@wiso.uni-erlangen.de

Abstract. Business processes have been widely accepted as the key factors in designing organisational structures. Tools and methods that support process design are limited to abstract design principles, general handbooks or rigid reference models and are therefore insufficient.

This paper provides a theoretical foundation for using reusable Process Building Blocks to design organisational processes. In this novel approach to Business Process Reengineering enterprises can design their processes by selecting, combining and customising Reference Process Building Blocks (RPBs) provided by a library.

RPBs represent design patterns and Best Practices for the organisational design and IT-support of business processes.

Keywords. Business Process Reengineering, Business Process Modelling, Reference Models, Reference Process Building Blocks (RPBs), object-oriented Process Design, Reusability, RPB library, Coordination Code (COORD)

1 Introduction

1.1 State of the Art

Business processes have been widely accepted as the key factors in designing organisational structures. In the last years a large number of various methods for designing business processes have emerged, for example Business Process Redesign [Davenport 1990], Core Process Redesign [Kaplan 1991] and Business (Process) Reengineering [Hammer 1993, Hammer 1995], which have influenced the debate about modelling business processes. Most methods that guide the process design effort follow a basic scheme (Fig. 1): process design starts with the strategic assessment and identification of the core processes. The main focus is on the organisational design and the IT-support of the business processes followed by testing and fully implementing the design solutions and ends in the establishment of a continuous improvement process.

For the organisational design and IT-enablement of processes two pervasive approaches are currently in use:

- **Conventional approach** Starting point in the conventional approach is a thorough view of the as-is state of the process followed by an in depth analysis of the process. The diagnosis of the deficiencies in the process provides the basis for the organisational redesign and IT-enablement. Alternatively, the search for design solutions on a "greenfield site" without a detailed process analysis can create breakthrough results. Performing process design in this way can be compared with producing tailor-made goods which often leads to high costs, low productivity and disappointing results. The key problem is a lack of methods and tools providing reusable process know-how, so that in most cases process design starts "from scratch". Therefore a process designer has to rely on his intuition and experience.
- **Reference Process Model approach** The main idea of this approach is to provide process know-how that is manifested in the form of generalised Reference Process Models. This more convenient way to use proposed design patterns of processes that are defined as Reference Process Models attempts to diminish the insufficiency of tools for designing processes [IDS Prof. Scheer 1995]. Problems with mapping entire business processes as reference models are presented by the high complexity and limited insight reflected by the process model as well as the missing capability to show alternatives within the model. Another critical aspect of Reference Process Models is that the organisational process and IT-support are not modelled simultaneously. In

addition, Reference Process Models concentrate on selected branches which narrows the range of use.

Two further approaches in designing business processes are currently being considered and have not been put to practice yet:

- **Skeleton approach** The Skeleton approach discusses aspects of supporting the reuse of software components. The idea of utilising macro structures which are refined according to the requirements of the enterprise is transferred to the field of process design. Depicting processes which rely on the Skeleton approach opens up a large range of design possibilities but still stands in need for supporting tools. This approach is similar to the idea of Bertram, who suggests using process templates which are adapted to the specific demands of enterprises [Bertram 1996: p. 93].

- **Reference Process Building Block approach** This novel approach focuses on improving the quality of process design by utilising reusable Reference Process Building Blocks (RPBs). RPBs represent design patterns and Best Practices for the organisational design and IT-support of business processes. With this innovative approach enterprises can design individual process models by selecting, combining and customising RPBs provided by an electronic library. A fundamental framework for RPBs is outlined in this paper.

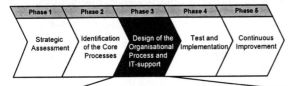

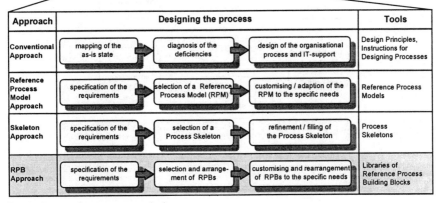

Fig. 1: Approaches to process design

1.2 Basic Approach

The basic approach consists in designing business processes by combining reusable, abstract Process Building Blocks that we refer to as *Reference Process Building Blocks (RPBs)*. The main characteristic of this methodology is the simultaneous design of process and IT-support, which is accomplished by embodying both aspects coexistent in the form of integrated process solutions and design patterns. Further, RPBs represent Best Practices of how processes can be performed and are modelled on varying levels of abstraction. The inference of RPBs on a problem and enterprise level is intended on the basis of the abstracted design ideas.

The modelling of RPBs focuses the utilisation of innovative IT as enabler together with the generation of new coordination mechanisms for combining RPBs to create innovative process solutions. RPBs that manifest the process solutions are stored in electronic libraries.

The business processes are designed in an innovative way by selecting and combining alternative RPBs provided by the library. In a next step, the modelled "process chain" composed of the combined RPBs is customised and adapted to the specific demands of the enterprise supported by object-oriented mechanisms.

RPB FRAMEWORK

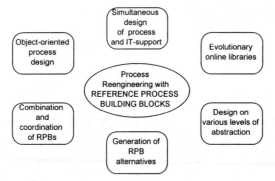

Fig. 2: RPB framework

1.3 Goals

In general, the RPB framework intends the realisation of the following goals:
* Higher quality of process design efforts,
* Reduction of time and costs.

This general intention receives significance in regard to the causes of insufficient design efforts. Detailed and in depth examinations and documentations of the as-is state of the process are major causes explaining why design efforts lack success.[1] Accordingly, they are time-consuming and lead to high costs. As a consequence, too little time is spent on designing an innovative new process that will lead to low quality design efforts. The RPB approach focuses on the essentials in designing business processes by offering easy-to-use process solutions that will allow the user to start right off.

The RPB approach can be distinguished from a large number of pseudo-integrated approaches that offer an entire phase model starting with a strategic assessment and resulting in the development of application software. In contrast, the development of the RPB methodology derives from a perspective of modelling business processes. This is due to the realisation that focusing too strongly on developing applications hinders the process design effort. As a result, the RPB methodology is able to meet the requirements of designing business processes in an effective and successful way.

1.4 Related Work

Within the research field of *Enterprise Modelling* reference models intend to support the mapping, explication and design of business processes, data models, activity diagrams and organisational structures. Scheer recognises their primary value in manifesting approved research results in structured models that are intended for reuse [Scheer 1994b]. In general, one can distinguish between *enterprise reference models, software-specific reference models,* e.g. SAP R/3 [Keller 1994: p. 6-18] and *reference models for developing applications,* e.g. CC RIM-Referenzmodell [Gutzwiller 1994]. The RPB methodology concentrates on exploring a new way of designing organisational processes with reference models within the wider field of enterprise modelling. The RPB framework has been influenced by the following related research work.
* Within the research project *CIM-OSA* (Computer Integrated Manufacturing Open System Architecture), initiated by ESPRIT, a reference framework for planning, designing and implementing CIM-Systems has been developed. CIM-

---

[1] "... perhaps the most commonly made error in reengineering is to spend far too much time analysing existing processes" [Hammer 1995: p. 17].

OSA enables the user to derive partial reference models for specific branch and enterprise types on the basis of pre-defined generic archetypes. Subsequently, the partial reference models can be refined to enterprise-specific models [Jorysz 1990a: p. 144-156; Jorysz 1990b: p. 157-166].

- The project *Toward a Handbook of Organisational Processes* [Malone 1993] is centred at the Massachusetts Institute of Technology, Centre for Coordination Science and has published its first results in 1993. As principle investigator Malone states, "the project includes (1) collecting examples of how different organisations perform similar processes, and (2) representing these examples in an online process handbook which includes the relative advantages of the alternatives" [Malone 1993: p. 1]. The Process Handbook is influenced by corresponding projects in cooperation with other universities, mainly the research works *Grammatical Models of Organisational Processes* [Pentland 1994b], *A Taxonomy of Organisational Dependencies and Coordination Mechanisms* [Crowston 1994] and *Applying Specialisation to Process Models* [Wyner 1995].

- The company IDS Prof. Scheer has expanded *Reference Process Models*, which concentrate on specific branches and have become more and more popular [IDS Prof. Scheer 1995]. They focus on process models in terms of the *Event-Driven Process Chain (EPC)*[2] integrating their relation to corresponding data, functional and organisational models.

The approach to designing processes with the help of reusable Process Building Blocks has merely been discussed. The idea of defining processes as standardised building blocks is initially founded within the RPB framework. However, many authors use similar notions, e.g. process components, core activities or sub-processes, when discussing the topic of process design. Becker, for example, considers a way to design processes by selecting process schemes and structure building blocks from a library and refining them according to specific needs [Becker 1995: p. 441].

---

[2] An EPC is a representation form for process modelling of the category *Event diagrams* [Keller 1992].

2 Simultaneous Design of Process and IT-Support

2.1 Definition and Key Features of RPBs

We define Reference Process Building Blocks as business processes and business process components which represent design patterns and process solutions for the organisational design and IT-support of processes. The structure and content of RPBs have been standardised in order to improve the design of processes.[3] The key features of RPBs are shown below.

Feature	Description
Character of reference	The main feature of an RPB is the representation of "best in practice" solutions for how a process can be performed integrating the organisational design and IT-support.
Reusability	The value of an RPB is based upon its ability to be reused in order to generate alternative and innovative process solutions.
Standardised structure and interfaces	The requirement to build on a standardised structure is fulfilled by defining standards for the layout and properties of RPB. The interfaces of RPBs are standardised by the use of Antecedent and Consequent Events from the process notation of the EPC.
Hierarchical representation	Using decomposition and specialisation (generation of alternative RPBs) leads to the development of RPB hierarchies that represent Process Building Blocks on various levels of abstraction.
Abstraction and taxonomy	RPBs represent abstract process solutions by providing only the information that is in fact necessary for the process design effort to ease and widen the use of RPBs. The development of RPBs relies on modelling process types on varying levels of abstraction.
Universal validity	RPBs are provided for universal use in a wide variety of enterprises and consequently do not include information related to specific enterprises. Universally valid RPBs can be arranged and customised in order to generate enterprise-specific process chains.
Modularised development	RPBs are developed based on an input-output related and abstract point of view and represent process modules that produce defined results. Thus, each RPB can be regarded as discrete.
Easy retrieval	Definite identification codes and the development of hierarchical models of RPBs guarantee an easy search and retrieval.

Table 1: Key features of RPBs

---

[3] The object-oriented view of a business process leads us to dissolve the difference between the notion of a business process, a function and an activity.

2.2 Interdependencies of Process Design and IT-Support

One of the most common reasons why process design fails lies in the primary search for organisational solutions and the subsequent adjustment of technological opportunities. Instead, "thinking about information technology should be in terms of how it supports new or redesigned business processes, rather than business functions or other organisational entities" [Davenport 1990: p. 12]. Due to the strong interdependencies of process design and IT-support, the view of a simultaneous design effort has been recommended, e.g. as used in the MIASOI-Method [Petrovic 1994].

RPBs are intended for easy and fast application. Moreover, the user has to succeed in designing the organisational process and the IT-support in the best possible way by relying only on the integrated process solutions and patterns given by the RPBs. Therefore, RPBs are centred around easy-to-understand and accurately detailed process solutions facilitating immediate use. However, modelling with RPBs is not expected to provide a framework for the development and implementation of application systems. To meet these requirements the necessary information is provided in the form of attributes embodied in the fundamental RPB Model as described in the next section.

2.3 RPB Model

Every RPB consists of a standardised structure with defined basic properties (Fig. 3). These basic properties make up the fundamental attribute classes of the *object RPB*. Each attribute class includes single attributes that describe the object RPB in order to meet the needs of the process design effort. These single attributes are the key elements of the RPB Model and will be referred to as *basic attributes*.

Coordination Attributes

The attribute class *Identification* includes a definite identifier as well as a description of the RPB. The *Coordination Code (COORD)* describes the decomposed RPB chain in semi-formal terms. In this way the COORD provides a semi-formal representation of the relationships between the various RPB layers that constitute the RPB library, thus supporting the navigation inside the RPB library. The COORD formalism is based on an easy grammar (see section 4.4).

Antecedent and Consequent Events represent the interfaces of an RPB and are modelled for the purpose of connecting RPBs in a proper way.

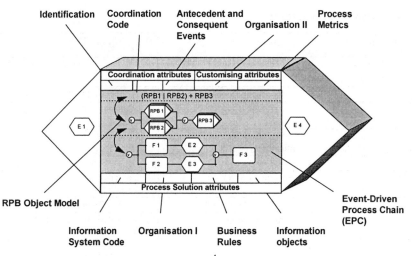

Fig. 3: Reference Process Building Block (RPB) [4]

Attribute class	Attribute	Description
Identification	identifier	short definite key (primary key)
	RPB designation	activity and object-reference of the RPB
	process solution	description of the key elements of the process solution in free-form text
	version number	current version number
Coordination Code (COORD)	COORD	process description in semi-formal terms
Antecedent and Consequent Events	Antecedent Event	defined starting event of an RPB
	Consequent Event	defined ending event of an RPB

Table 2: Coordination attributes of an RPB

---

[4] The semi-formal COORD can be transformed to the more perceivable form of the RPB Notation or the Event-Driven Process Chain. Thus, the business process can be expressed in alternative ways, in semi-formal terms of the COORD, as a chain of RPB objects or in terms of the Event-Driven Process Chain.

Process Solution Attributes

Attribute class	Attribute	Description / Example
Information System Code (ISC)	application systems	e.g. Workflow Management System
	help / support systems	e.g. electronic time scheduler
	level of IT-support	manual | partly automated | automated
	integration to other applications	e.g. customer information system
	databases	e.g. customer database
	information storage	e.g. tape | hard disc | disc | chip-card | CD-ROM | paper
	hardware components	e.g. touch-activated screen
	carrier techniques	e.g. wired | wireless | paper transport | verbal
	communication systems	e.g. telephone | fax | email | video-conferencing
Organisation I (ORG I)	organisational category	e.g. customer | supplier
	division of labour	case worker | case team
Business Rules	Business Rules	e.g. yield-management
Information Objects (IO)	IO-designation	description of the IO, e.g. order data
	relationship of IO and activity	produces | uses | modifies | requires

Table 3: Process Solution attributes of an RPB[5]

The *Information System Code (ISC)* shows the IT-systems and the way they support the RPBs. The choice and application of organisational units that will be responsible for performing the RPBs are stored in the attribute class *Organisation I*. The key ideas for process solutions can often be found within methods or principles. These ideas can seem meaningless in isolation, but can, however, create breakthrough process solutions when used to generate new process scenarios. We will refer to these sort of key ideas as *Business Rules*. Lastly, the information objects needed for the process design effort belong to the attribute class *Information Objects*.

Customising Attributes

When customising arranged RPB chains, empty attribute values can be filled, and attribute values can be modified. In addition, new attributes may be specified which can be filled during the customising process. An example of these Customising attributes are attributes which specify the individual organisation of the process, e.g. the name of the process owner. Another example are process

---

[5] In the table the straight line "|" can be read as single or multiple choice of possible values of the basic attributes.

metrics which can be filled on an on-going basis to support process controlling and continuous improvement efforts.

2.4 RPB Views

Developing and applying RPBs includes various steps that focus on specific parts of the available information pool (RPB attributes). Thus, the RPB framework supports the use of predefined views that provide the user with selected information in correspondence with her/his current design focus, thereby temporarily reducing the overall information load. One can choose from different *Process Solution views, Coordination views* and *Customising views*, which provide filtered information depending on the current task. Moreover, self-made views can be set up to accomplish individual efforts relying on specific attributes.

2.5 RPB Categories

The RPB framework distinguishes between two main categories of RPBs:
- **Generic RPBs** These RPBs are modelled on a high level of abstraction and do not relate to any specific domain (e.g. specific process types or branches). They include generic process solutions and design patterns and offer the highest level of reusability within all categories of the RPBs. Examples, on an elementary level, are "collecting information" and "transmitting information".
- **Domain RPBs** Domain RPBs refer to a specific field of interest, e.g. a certain process type or branch. An example for a Domain RPB is "acquiring billing data through electronic voice recognition".

Additionally, Generic and Domain RPBs are separated into General RPBs and Process Solution RPBs.
- **General RPBs** General RPBs are specified only by their Coordination attributes. By combining General RPBs, templates are created which constitute the structure for the building of RPB libraries.
- **Process Solution RPBs** Process Solution RPBs add a process solution to the Coordination attributes, which is expressed in their Process Solution attributes as ISC, Business Rules etc.. They are of major importance for the process design effort as they embody the essential process know-how.

Moreover, General RPBs as well as Process Solution RPBs can be classified according to the object they refer to, i.e. the object they basically deal with, e.g. the object "customer information". The *object-reference* determines the core activity along with the process solution of an RPB.

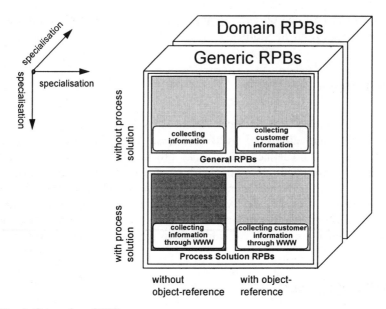

Fig. 4: Categories of RPBs

Of primary interest within the RPB framework are *Generic RPBs* which include process solutions (*Process Solution RPBs*) and do not refer to a specific object (*without object-reference*). They embody key ideas and design solutions that can be applied within a wide range of use. Moreover, they are an important basis upon which Domain RPBs can be developed. Generic RPBs are provided on various levels of abstraction relying on a basis taxonomy. The following example will illustrate the notion of Generic RPBs with process solution and without object-reference. On an elementary level of abstraction the Generic RPB "collecting information" is offered along with process solutions as:
- manual collection of information by using touch-activated screens
- partly automatic collection of information with product code scanners
- automatic collection of information through the use of smart-cards

Specialisation and inheritance mechanisms support the development of Domain RPBs, e.g. the RPB "collecting orders with a touch-activated screen". This process solution is currently being applied by the mail-order retailer Otto Versand who supplies BP stations with touch-activated screens which allow customers to fill in their orders.

3 Object-oriented Process Design

3.1 Object-oriented Notion of a Process

In the RPB methodology principles and methods of *object-oriented design* are exploited for process modelling. The numerous object-oriented approaches [Coad 1991, Booch 1991] consider static items (e.g. customer, invoice) as *objects* and *classes* and describe them by *attributes* and *methods*.

In contrast to the conventional notion of object-oriented design, the RPB approach defines dynamic processes and process components as *objects*. In the *object scheme* the basic attributes of an RPB (e.g. COORD, carrier techniques) correspond to the attributes and methods of object-oriented design. For instance, the class *handling orders (alternative A)* contains activities such as *receiving orders, checking orders* and, for the IT-support, a *workflow management system*.

This point of view permits the transfer of principles and methods of object-oriented design to the RPB approach. As a result the development of *RPB class hierarchies* is supported. In another aspect, by applying *inheritance mechanisms* RPB attributes (e.g. application systems) can be inherited to subordinated RPB objects. This procedure is the key to generating alternative and new RPBs and it especially eases the customising of RPBs in regard to the individual circumstances of enterprises. The adaptation of RPBs to individual circumstances corresponds with the *instantiation of objects*. For example, the RPB *paying an invoice via Internet at the company Jackson and Co.* is an instance of the RPB *paying an invoice via Internet*.

The definition of an RPB in object-oriented terms dissolves the differentiation between business processes, process components and activities. All processes, irrespective of their extension and content, are defined as *RPBs* or *RPB objects*. As a consequence, RPB objects are the only components for modelling processes.

3.2 Decomposition and Specialisation

Defining processes as objects enables one to adopt decomposition and specialisation mechanisms from the object-oriented design.

Using *decomposition* leads to decomposed RPBs, which represent disjunct objects, on a more detailed level of abstraction. As a result of applying the decomposition mechanism RPBs are portrayed on various levels of abstraction. The decomposition ends on a level of abstraction that describes atomic activities.

Representing RPBs on various levels of abstraction contributes to the systematic development of RPB hierarchies and reflects the layered quality of processes.

A top-down approach to process design, as suggested by Michael Hammer [Hammer 1993], is facilitated by progressive decomposition. For instance, the RPB *handling orders* is decomposed into the disjunct RPBs *receiving orders, checking orders, confirming orders* and *fulfilling orders*.

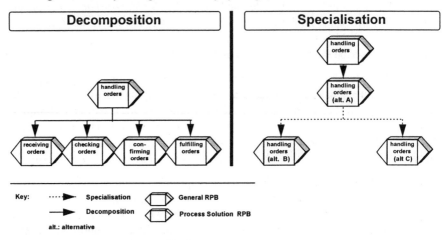

Fig. 5: Decomposition and specialisation

In contrast to the notion of "function trees" which depict static functions, RPB hierarchies describe dynamic processes. To formalise the sequence in which the decomposed RPBs are combined the RPB methodology suggests using the *Coordination Code (COORD)*, a process language in semi-formal terms (see section 4.4).

Specialisation leads to new classes of the object RPB. In general, a new RPB is created by inheriting the attributes (e.g. the attributes application systems or COORD) of an existing RPB to a new RPB and then modifying / adapting selected attributes of the created new RPB. As illustrated in Fig. 5, alternatives of the RPB *handling orders* have been developed. At first, the Process Solution RPB *handling orders (alternative A)* is derived from the General RPB *handling orders* by applying specialisation. Because of having started with a General RPB (RPB without a process solution), additional attributes of the RPB *handling orders (alternative A)* have to be specified. In a second step, the RPB *handling orders (alternative A)* is specialised to the alternatives B and C through the use of inheritance mechanisms. At last the attributes of alternatives B and C are refined to meet the individual requirements that were set up as business targets for designing the new alternatives. In contrast to decomposition, which leads to disjunct RPB, specialisation provides *non-disjunct* RPBs.

By applying specialisation, new and innovative alternatives of RPBs that represent innovative process solutions can be easily generated. Also, this easy way to generate alternatives supports the development of RPBs. However, the great advantages of specialisation in the process design effort become effective in designing and customising the processes. Especially the gap between the universal RPBs and the individual demands of the enterprise can be reduced during the customising process by systematically adopting decomposition and specialisation. In this way the customising process mainly involves refining the RPB attributes step-by-step to meet the individual needs of enterprises. Furthermore, process modelling becomes independent of the complexity of enterprises.

3.3 Transformation of the RPB Model into EPC

The RPB methodology includes a method for transforming an RPB Model into EPC. The EPC is not an additional attribute of RPBs, only a graphical representation of the attribute COORD. The COORD describes a process (here as a chain of RPBs) in semi-formal terms, thereby reducing their complexity and supporting the combination of RPBs.

The transformation of an RPB Model into EPC requires superscribing the overlapping *Antecedent* and *Consequent Events*. For this, rules have to be defined as *the Consequent Event superscribes the Antecedent Event of the following RPB*. The outcome of the transformation is the *EPC* that represents - in contrast to the semi-formal COORD syntax and the RPB Notation - the coordination order of the RPBs in a more comprehensible form.

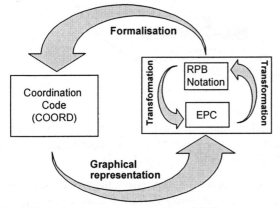

Fig. 6: Transforming the representation of RPBs

4 Combination and Coordination of RPBs

4.1 Dynamic Application Cycle

Process design can be understood as a dynamic coordination process of all components of a business to support the strategically objectives. The dynamic application cycle reflects this philosophy and defines a framework that includes different coordination tasks. Rapidly changing conditions are the trigger for the cycle, which embodies the following steps.

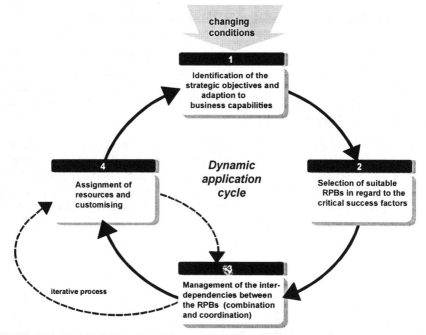

Fig. 7: Dynamic application cycle

❶ Rapidly changing markets and customer demands require a regular analysis of business strategy. This includes the reconsideration and identification of the strategic objectives and their influence on business capabilities.

❷ The critical success factors for the process design effort, the process objectives and process metrics are derived from the newly defined business strategy. The critical success factors (e.g. customer service) and process objectives are the key elements for selecting suitable RPBs.

❸ Next, the selected RPBs are combined and coordinated in order to meet the critical success factors in the best possible way. A key challenge in coordinating RPBs to create efficient processes is to match the interdependencies of the RPBs (e.g. different IT-support, interfaces). This third step is called *the management of the dependencies between the RPBs* which is supported by mechanisms for coordinating RPBs.

❹ Finally, the coordinated RPBs have to be adapted to the individual requirements of the business situation (*customising*). The customising process is centred around the main mechanisms of decomposition and specialisation, but may require a rearranging of the RPBs.

Steps three and four have a mutual impact on each other and cannot be separated, but are proceeded in an iterative cycle. For instance, changing the IT-support of an RPB might require a new coordination mechanism.

4.2 Combination of Reference Process Building Blocks

The *Combination of RPBs* is understood as connecting single RPBs to an entire process chain or process model. RPBs can be combined sequentially as well as in terms of either an AND-connection, an OR-connection or an XOR-connection.

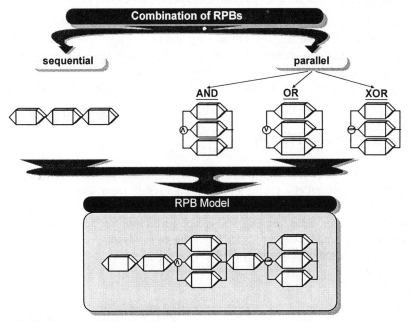

Fig. 8: Combination of RPBs

In contrast to the terminus *combination*, which focuses on syntactical rules for the act of combining (e.g. rules for connecting interfaces), the terminus *coordination* illuminates semantically issues. This corresponds to Malone, who

282

define coordination as "...the act of managing interdependencies between activities" [Malone 1990: p. 12].

4.3 Coordination of Reference Process Building Blocks

The major effort within the dynamic application cycle is to manage the dependencies between the RPBs (step three), which is influenced by the customising process. The effort lies in coordinating the RPBs flexibly to meet the process objectives and success factors.

Recent research in the field of coordination has been accomplished by Malone and Crowston, who have concentrated on developing an *interdisciplinary coordination theory* [Malone 1990] and by Pentland who has explored coordination possibilities within processes relying on *process grammars* [Pentland 1994]. Generally, other authors reduce coordination tasks to the act of allocating resources, i.e. the assignment of rare resources to activities. However, there is a lack of methods which systematically support the exploration of coordination possibilities that can be adopted to related or specific coordination problems.

The RPB methodology requires coordination mechanisms that allow the user to combine and coordinate the RPBs in an easy and proper way. These coordination mechanisms could be represented in reusable order relations of RPBs chains. In this research project a *coordination matrix* was developed that allows the coordination possibilities to be evaluated systematically [Richartz 1995: p. 25-28].

4.4 Coordination Code

The COORD is a semi-formal notation for representing RPBs. It describes the dependencies and relations of the RPBs that result from the decomposition mechanism. It depicts the RPBs of the next subordinate level and their chronological order. Thus, the COORD is a powerful tool for modelling processes on various, integrated levels of abstraction. The great advantages lie in facilitating the combination and coordination of RPBs.

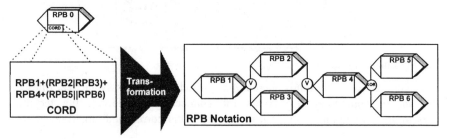

Fig. 9: Coordination Code

Rules
- The COORD is assigned to the RPBs that is decomposed into subordinated RPBs. Consequently, to avoid redundancy, the COORD does not provide information about the upper, decomposed RPB.
- The COORD only describes the relation between two hierarchy layers.
- The *Antecedent* and *Consequent Events* are stored outside the COORD in other RPB attributes. However, the notation of the COORD can also be used to describe complex structures of events.

Major functions of the COORD:
- **"Assembling" RPB chains** The COORD supports the generation of RPB chains. The order of RPBs documented in the COORD can easily be modified by the user.
- **Representing the universal structure of an RPB library** The COORD can be used for describing the structure of an RPB library by aggregating the single COORD of all RPBs.
- **Representing the entire RPB library (including alternatives of RPBs)** An overall COORD of a library can be generated by combining the COORD of all RPBs together with all alternatives of RPBs that are fixed in the Table of Alternatives (see section 5.2).
- **Storing RPBs without redundancy** The RPB methodology intends to avoid redundant storage on the basis of a pointer system. Accordingly, all RPBs include a pointer that refers to the definite physical location, so that every RPB can be reused in numerous created RPB chains but is stored only once.

5 Generating RPB Alternatives

Conventional methods for designing processes, especially Reference Process Models, describe *General Processes* that embrace only a small part of all possible ways to perform the processes. Therefore, they narrow the range of use in designing new and individual processes. Due to this common deficiency, a key element of the RPB methodology is to generate RPB alternatives in order to enhance the flexibility and individuality for the process design.

5.1 Applying Specialisation to RPBs

Specialised RPBs are generated by adding information to a given RPB which leads to a higher level of information depth of the specialised RPB. For instance, a specialised RPB can be generated by specialising a *General RPB* to a *Process Solution RPB*.

284

The main characteristic of *RPB alternatives* or *alternative RPB relations* is that they are located on the same level of information depth. RPB alternatives can therefore be substituted by other appropriate RPB alternatives at any time. Moreover, RPB alternatives can be generated in two ways: on the one hand by applying specialisation to RPBs while increasing the depth of information and, on the other hand by modifying existing RPB alternatives while remaining on the same level of information.

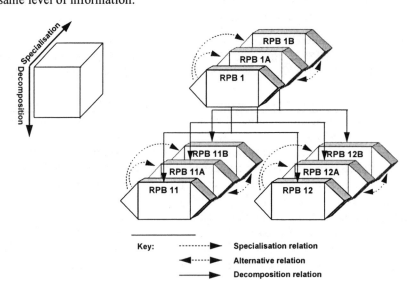

Fig. 10: Applying specialisation to RPBs

The idea of providing the user with RPB alternatives gains importance considering the easy way of how RPB alternatives can be generated through the support of inheritance mechanisms. Generating an alternative RPB starts by transferring the RPB attributes to a new RPB (inheritance) and ends in modifying and adapting the attributes within the created new RPB alternative.

5.2 Table of Alternatives

In contrast to the Coordination Code, which represents the decomposition of RPBs the *Table of Alternatives* (*TALT*) specifies specialisation and alternative relations between RPBs.

The TALT depicts the specialised RPBs and RPB alternatives of an RPB in a systematic way. This additional tool works complementary to the COORD which is unable to describe the specialisation and alternative relations of RPBs.

6 Designing on Various Levels of Abstraction

6.1 Deficiencies in Focusing one Level of Abstraction

To a large extent, existing Reference Process Models are described on *one* level of abstraction, which results in large and complex process models. Even when processes are designed on a medium level of abstraction, the overview of the process is reduced. This can lead to a misunderstanding of the process and enhances the risk of failing to notice process deficiencies. What is more, designing processes on one level of abstraction leads to the problem of finding the "right" level without losing too much information [Marent 1995: p. 312]. Yet, the "right" level of abstraction cannot be defined universally, e.g. as Reference Process Models implicitly suggest. Instead, the appropriate level depends on the individual conditions of processes. Designing on a too detailed level of abstraction may limit the reusability of the reference model, while designing on a too high level may simplify the model too far and risk the loss of important information.

The RPB methodology avoids the problem of having to define the adequate level of abstraction for the design effort by providing RPBs on increasingly detailed levels of abstraction. This key characteristic facilitates the development of RPB hierarchies. Moreover, designing on various levels of abstraction supports a top-down approach in the design effort allowing the user to structure and refine the processes step-by-step. An additional effect is the wide range of usability that is opened up by offering RPBs on various levels of abstraction. RPBs can be flexibly adapted to different kinds of individual processes, whether they be complex processes of large enterprises or simple processes of small companies.

6.2 RPB Hierarchies

RPB hierarchies are constructed in order to represent the various levels of abstraction. They show the multi-layered structure of RPBs. Creating an RPB hierarchy is based on the application of the decomposition mechanism.

In addition, by integrating specialised RPBs and RPB alternatives, more complex hierarchies can be created. These hierarchies are able to illustrate all relations of RPBs of an entire RPB library.

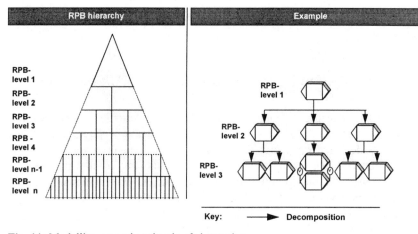

Fig. 11: Modelling on various levels of abstraction

7 Evolutionary Online RPB Libraries

7.1 Development and Evolutionary Extension of RPB Libraries

The RPB methodology provides an easy-to-use way of evaluating Best Practices and process solutions and representing them in RPB libraries in a structured and uniform way. These RPB libraries have the capacity to include all RPBs of a predefined domain and can be seen as knowledge bases for designing processes.

Designing processes with the RPB methodology is only satisfactory when a sufficient number of RPBs is available. Consequently, one main objective within the RPB framework is to extend the libraries relying on an effective way of developing RPBs.

The evolutionary development and extension of RPB libraries is managed by following an iterative cycle (see Fig. 12). An important element in designing processes with RPBs is the ability to generate new innovative processes. The exploration of innovative RPBs plays a main role within the iterative cycle. The primary question is "how can the process be designed in a different, new way?". The starting point is to arrange and adapt RPBs in all possible and unusual ways. Then new process solutions can easily be revealed by following a systematic method that builds on the three major interdependent factors of coordination, information technology and organisational process.

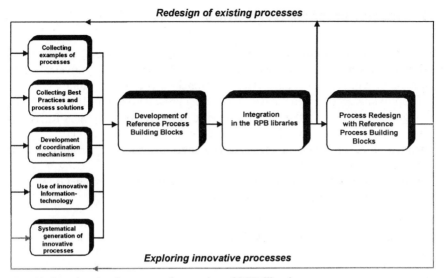

Fig. 12: Iterative development and extension of RPB libraries

7.2 Online RPB Libraries

The development of RPB libraries leads to a large amount of RPBs on different levels of abstraction and involves decomposition, specialisation and alternative relations. Moreover, each RPB is described by its basic attributes. To deal with this load of interdependent information requires an appropriate support of software tools and databases.

For this reason RPBs will be stored in electronic libraries e.g. represented by object-oriented databases. These *online libraries* will help to handle the information load and will provide fast information retrieval. Also, an online library is the key to succeed in applying methods as specialisation and mechanisms as inheritance.

A software tool will be important to support the RPB methodology in many ways, such as enabling an easy and fast retrieval and supporting the arrangement and the coordination of RPBs as well as the customising process and the generation of RPB alternatives.

7.3 Categories of RPB Libraries

The RPB methodology provides several categories of RPB libraries:

- **Generic library** This library contains RPBs which are universal and independent of individual businesses and branches. The Generic library consists of approximately 300 Generic RPBs including process solutions. It will provide a high reusability for different purposes, such as developing Domain RPBs for other Branch and Business libraries which are explained below.
- **Branch libraries** RPBs of this category meet the specific requirements of branches (e.g. financial services).
- **Business libraries** Business libraries are developed by adapting Generic or Branch RPBs to the needs of individual businesses.

As mentioned above the main function of the Generic library is to support the development of Branch and Business libraries.

8 Conclusion

The RPB framework intends to improve the design of organisational processes and IT-support. The innovative approach is to provide RPBs which represent integrated Process Solutions. This enables the immediate and easy use of Best Practices in designing business processes, which leads to lower costs, less time consumption and a higher quality of the process design effort. The RPB methodology can be of fundamental help for all who rely on efficient tools for designing processes, e.g. consulting firms. Furthermore, the method can influence the research in designing organisational structures.

Process design with RPBs has not yet been approved in practical use. However, on a theoretical level, using RPBs promises essential advantages in contrast to conventional tools. The possibility of systematically combining and customing RPB chains by selecting from alternative RPBs offers a high flexibility in finding process solutions that will suit the individual needs best, e.g. in contrast to the considerable inflexible nature of fixed Reference Process Models. Another difference to more common tools is that process solutions of RPB integrate the interdependencies of process and IT-support. Moreover, RPBs have the capability to express the nested, layered quality that characterises many kinds of organisational processes which widens the range of use.

The easy way of arranging and rearranging RPBs that embody ideas for the innovative utilisation of IT as well as novel coordination mechanisms opens up a wide range of ways to explore innovative process solutions. By systematically evaluating new ways of performing processes with the RPB methodology, enterprises can gain substantial competitive advantages.

RPBs comprise process solutions that reveal the key ideas in a compact and understandable way and are incorporated in a standardised structure. These aspects facilitate an easy development and trade of RPB. Looking in the future you can imagine markets for RPBs established within various branches. In this vision RPBs can be thought of being traded among different branches as well as being sold from professional RPB designers, who refer to RPBs as process know-how.

References

Becker, J., Rosemann, M. and Schütte, R. (1995) Grundsätze ordnungsgemäßer Modellierung. *Wirtschaftsinformatik*, 37, p. 435-445.

Bertram, M. (1996) Das Unternehmensmodell als Basis der Wiederverwendung bei der Geschäftsprozeßmodellierung, in *Geschäftsprozeßmodellierung und Workflow-Management* (ed. Vossen, G. and Becker J.), Bonn, p. 81-100.

Booch, G. (1991) *Object-Oriented Design.* Redwood City.

Coad, P. and Yourdon, E. (1991) *Object-Oriented Analysis.* Englewood Cliffs.

Crowston, K. (1994) *A Taxonomy of Organisational Dependencies and Coordination Mechanisms.* Working Paper, The University of Michigan, School of Business Administration, Tappan.

Davenport, T.H. and Short, J.E. (1990) The New Industrial Engineering: Information Technology and Business Process Redesign. *Sloan Management Review*, 4, p. 11-27.

Gutzwiller, T.A. (1994) *Das CC RIM-Referenzmodell für den Entwurf von betrieblichen, transaktionsorientierten Informationssystemen.* Heidelberg.

Hammer, M. and Champy, J. (1993) *Reengineering the Corporation.* New York.

Hammer, M. and Stanton, S.A. (1995) *The Reengineering Revolution - A Handbook.* New York.

IDS Prof. Scheer (ed.) (1995) *ARIS-Methodenhandbuch.* Buch 5, Release 3.0, Saarbrücken.

Jorysz, H.R. and Vernadat, F.B. (1990a) CIMOSA Part 1: total enterprise modelling and function view. *International Journal of Computer Integrated Manufacturing*, 3-4, p. 144-156.

Jorysz, H.R. and Vernadat, F.B. (1990b) CIMOSA Part 2: Information view. *International Journal of Computer Integrated Manufacturing*, 3-4, 157-167.

Kaplan, R.B. and Murdock, L. (1990) Core Process Redesign. *The McKinsey Quarterly*, 2.

Keller, G. and Meinhardt, St. (1994) *SAP R/3-Analyzer: Optimierung von Geschäftsprozessen auf Basis des R/3-Referenzmodells.* SAP AG (ed.), Walldorf.

Keller, G., Nüttgens, M. and Scheer, A.W. (1992) *Semantische Prozeßmodellierung auf der Basis Ereignisgesteuerter Prozeßketten (EPK).* Veröffentlichungen des Instituts für Wirtschaftsinformatik, Heft 89, Saarbrücken.

Malone, T.W. and Crowston, K. (1990) *Toward an Interdisciplinary Theory of Coordination.* Working Paper, Center for Coordination Science, Sloan School of Management, Massachusetts Institute of Technology, Cambridge.

Malone, T.W., Crowston, K., Lee, J. and Pentland B. (1993) Tools for Inventing Organizations: Towards a Handbook of Organisational Processes, in: *Proceedings: Second Workshop on Enabling Technologies Infrastructure for Collaborative Enterprises,* Los Almitos, p. 72-82.

Marent, C. (1995) Branchenspezifische Referenzmodelle für betriebswirtschaftliche IV-Anwendungsbereiche. *Wirtschaftsinformatik*, 37, p. 303-313.

Pentland, B.T. (1994) *Process Grammars: A Generative Approach to Process Redesign.* Working Paper, University of California, Los Angeles.

Pentland, B.T. (1994b) *Grammatical Models of Organisational Processes.* Working Paper, University of California, Los Angeles.

Petrovic, O. (1994) MIASOI: Ein Modell der iterativen Abstimmung von Strategie, Organisation und Informationstechnologie, in *Geschäftsprozeßmanagement* (ed. Krickl, O.C.), Heidelberg.

Richartz, P. (1995) *Vorgehensmodell und Modellierungsmethoden für den Einsatz von Referenzprozeßbausteinen zur Geschäftsprozeßoptimierung.* Thesis, Universität Erlangen-Nürnberg, Lehrstuhl Wirtschaftsinformatik II, Nürnberg.

Scheer, A.-W. (1994) *Wirtschaftsinformatik - Referenzmodelle für industrielle Geschäftsprozesse.* Berlin et. al.

Scheer, A.-W. (1994b) Geleitwort, in: *Referenzdatenmodelle: Grundlagen effizienter Datenmodellierung* (ed. Hars, A.), Wiesbaden.

Wyner, George M. and Lee, J. (1995): *Applying Specialization to Process Models.* Working Paper, Center for Coordination Science, Sloan School of Management, Massachusetts Institute of Technology, Cambridge.

An Object-Oriented and Business Process-Based Meta Model of an Architecture for Management Support Systems

Anett Mehler-Bicher

EUROPEAN BUSINESS SCHOOL (ebs), Schloß Reichartshausen, D 65375 Oestrich-Winkel, anett.bicher@ebs.de

Keywords. management support system (MSS), architecture model for MSS, business process-orientation, object-orientation, business process modeling, Semantic Object Model (SOM), meta model

1 Introduction

The current enterprise environment is especially characterized by changes. Reasons are for example the constantly progressive change from supplier- to buyer-markets, the increasing internationalism of the markets and the distinct customers' demands forcing the enterprises to an intensified market- and customer-orientation. As a consequence, the result from the economical activities of an enterprise depends more and more on its speed of response and its adaptability towards changing constraints – both internal and external. The enterprises come to meet this external change internally with more flexibility. A quick reaction and adaption towards changed constraints and a corresponding enterprise management require a permanent availability of complete, decision relevant and current information for the management. As a result, effective and efficient management support systems (MSS) especially supporting strategic decisions are called for by enterprises that want to retain their market position successfully. On account of their growing importance, the problems of existing MSS become more and more painfully apparent; more adequate MSS are required.

Obvious weaknesses of existing MSS are essential lacks of user-friendliness, flexibility, maintainability and integrability. Other, more fundamental deficiencies are the inadequate possibility to identify the managers' information needs and to transform them into the MSS. Most managers believe that existing MSS cannot satisfy their information needs. Conventional system development is the main reason for these lacks. Neither the complexity of management systems can be mastered nor the managerial tasks „structuring" or „managing" can be supported in a qualified manner.

It was realized that significant improvements may result as a consequence of careful analysis and re-design of existing technologies. Analyzing requires especially appropriate modeling methods. The deficiencies of traditional system development methods led to the concept of an information systems architecture (e.g. see (ESPRIT 1991, Hildebrandt 1992, Krcmar 1990, Österle, Brenner and Hilbers 1992, Olle 1989, Scheer 1992, Sowa and Zachman 1992, Zachman 1987). In order to master and control complexity, various modeling levels taking different views into account are provided. The relations between views and levels are represented in a comprehensive architecture. Thus, an architecture includes several related models using various description levels for representing different views. But a more detailed analysis of existing architectures points out that most of them are based on conventional system development ideas; as a result, they show corresponding deficiencies. The approaches either are little concrete or – under both economical and software-technical aspects – appropriate concepts˙ are missing. Integrative meta models to represent the comprehensive architecture as well as appropriate procedure models are missing. Especially views for modeling aspects concerning strategic business planning – of great relevance for the MSS development – are neglected. In addition to the concept of information system architecture, the key issues for today's companies and organizations are object-orientation and business process-orientation (e.g. see Ferstl and Sinz 1995, p. 209ff). The essential idea of object-oriented approaches to system development is to model structures which grant a most authentic description of the world of discourse. Business process-orientation aims to offer a dynamic viewpoint to enterprises. Business processes getting into the center of business analyses allow an entirely modeling of business systems. Both paradigms guarantee quality, innovation and flexibility and seem to offer a way to overcome the lacks of traditional system development.

Within the scope of this paper, the feasibility of business process-orientation and object-orientation concerning the development of management support systems (MSS) is discussed. Therefore an object-oriented, business process-based architecture model for MSS is presented. In the following section, first managerial tasks are analyzed and the term MSS is defined. After specifying the goals of MSS and a corresponding architecture, the synergies of object-orientation and business process-orientation for MSS development are stated. Then the meta model of the architecture for MSS is exposed. Finally the results are summarized and a short outlook is given.

2 Survey of the Architecture for MSS 0

2.1 Abstract Managerial Tasks „Managing" and „Structuring"

Management is defined as „*structuring* and *managing* purpose-oriented social systems" (Ulrich 1984, S. 99ff). Managing means fixing goals and specifying, initiating and controlling goal-directed activities of an enterprise. The management aspires to a goal-oriented influence of behavior in order to save viability long-sightedly. Structuring serves the creation of a business concept, i.e. a structure allowing an enterprise to make adaptations to changing conditions (Bleicher 1991, p. 35). The primary goal is planning, developing and maintaining an enterprise in a way that guarantees the ability of managing permanently. The aspects *services*, *feedback controlling targets* and *work-flows* are task-independent, management-important viewpoints to an enterprise. Using cybernetics ideas (e.g. see Malik 1992), the abstract managerial tasks can be described by basic coordination mechanisms. The *Semantic Object Model* (*SOM*) (e.g. see Ferstl and Sinz 1994a, Ferstl and Sinz 1994b, or Ferstl and Sinz 1995) – a comprehensive, both object-oriented and business process-based approach for analysis and design of business information systems and specification of business application systems – uses transactions for describing the basic coordination mechanisms

294

- negotiation principle: A business transaction between two objects is decomposed into a sequence of sub-transactions. During the initiating transaction, the objects learn to know each other and exchange information on deliverable services. Within the contracting transaction, both objects agree to a contract on the exchange of a service. The purpose of the enforcing transaction is to exchange the service between the objects (Ferstl and Sinz 1994b, p. 8).
- feedback control principle: A business object is decomposed into two sub-objects and two transactions, together establishing a feedback control loop. The controlling sub-object prescribes objectives or send messages to the controlled sub-object by a control transaction. A feedback transaction closes the feedback control loop by reporting to the controlling object (Ferstl and Sinz 1994b, p. 8).

Both principles grant especially the three management-important viewpoints mentioned above. In analogy to FERSTL/MANNMEUSEL's concept (Ferstl and Mannmeusel 1995, p. 450f), the feedback control principle is extended with the specialization *goal transaction*; this extension allows a more management-specific description of coordination mechanisms. In contrast to a control transaction, a goal transaction does not initiate a task execution at the controlled object, it only updates the formal objectives the controlled object has to consider within the scope of its decision space. Fig. 2-1 and 2-2 illustrates the different coordination mechanisms for management and structuring.

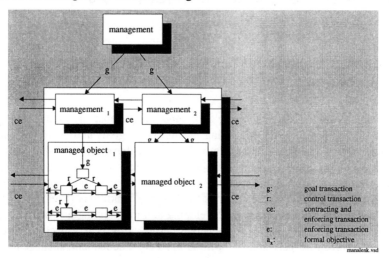

Fig. 2-1: Coordination Mechanisms for Managing.

In the case of managing (see Fig. 2-1), a manager defines goal transactions to the hierarchical directly subordinated objects; these objects are also controlling objects. Objects which are controlled by goal transactions receive targets to be considered in their decision spaces. An adequate proceeding for goal performance is specified by themselves personally. Thus, managing requires a *system of coope-*

rating business processes; concerning allowed structures of subordinated business processes see Ferstl and Mannmeusel 1995, p. 450ff.

In the scope of structuring activities (see Fig. 2-2), a manager initiates only *one business process* to all subordinated objects by the definition of a control transaction; the primary goal is that the specifications are regarded and executed in the intended way by all objects. Structuring specifications – both substantive goal-oriented and formal objective-directed – bind all subordinated objects. Whereas substantive goals are valid for all subordinated objects, formal objectives can differ object-specific. Thus, the formal objectives of specifications are explicitly handed over to directly subordinated objects.

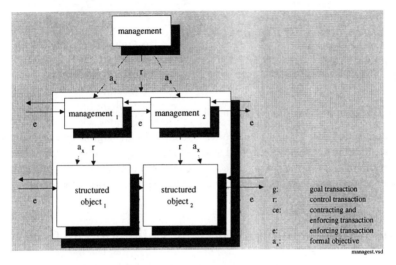

Fig. 2-2: Coordination Mechanisms for Structuring.

The difference between the managerial tasks managing and structuring becomes clearly obvious by analyzing the corresponding auxiliary controlled systems (see Fig. 2-3). The managing task requires a model of the concrete enterprise structures and their relationships. Business processes allow the description of an enterprise under both structure- and behavior-oriented viewpoints. That is the reason why the auxiliary controlling system corresponds to the system of business process models describing the controlling object's domain of task and competence. In the case of structuring, the controlling system is more complex. Structuring requires information and knowledge about the entire enterprise and its architecture, i.e. its principal structures and both static and dynamic relationships. Thus, the controlled system covers an – imaginary – management-specific meta model of the enterprise architecture and corresponding instances.

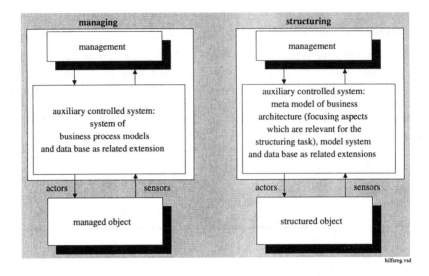

Fig. 2-3: Comparison of the Auxiliary Controlled Systems for Managing and Structuring.

An analysis of the auxiliary controlled systems for managing and structuring points out that the auxiliary controlled systems for managing can be interpreted as a sub-model of the auxiliary controlled systems for structuring. This relation allows to consider the managerial task managing as a specialization of structuring.

2.2 Management Support System: Definition and Goal

Although today new approaches are summarized under the term *management support system (MSS)*, a standardized definition does not exist. An MSS subsumes all – normally computer supported – means serving managers or other persons which take an essential part in the decision-making processes of an enterprise as an instrument for their managerial tasks. It represents a modular structured and integrated business information system. On the one hand, it has to guarantee the provision of decision representatives in the enterprise with managerial relevant, current and arbitrarily aggregated information from different in- and external data sources. On the other hand, it has to support the preparation and sub-processing in all phases of the decision-making process in order to support the management within its managerial tasks. Thus, the principal goal of an MSS is the entire support of the management within its managerial tasks in order to retain the viability of the enterprise.

Basing the definition of (Scott-Morton 1983, p. 5ff), an MSS subsumes all executive support systems (ESS) of an enterprise; an ESS represents a position-oriented combination of a decision support system (DSS) and an executive information system (EIS). Whereas ESS are „ ... focused on a manager's or group of managers' information needs across a range of important and relevant areas of concern ... [and] ... incorporate in a single system the data and the analytic tools needed to provide information support for a broad range of managerial processes and decisions" (Scott-Morton 1983, p. 5ff), DSS are „ ... focused on a specific decision or a specific class of decisions" (Scott-Morton 1983, S. 5ff). They also cover expert systems components and group decision support systems components. In contrast to DSS, EIS are characterized by an intuitive understandable structure; their focus is oriented towards the later phases of the decision-making processes in order to identify necessary need to act.

2.3 Architecture for MSS: Definition and Goals

In the scope of system development, information systems architectures are required in order to master and control complexity. Various modeling levels taking different views into account are provided. The relations between views and levels are represented in a comprehensive architecture model. An architecture model for MSS is a specific information systems architecture because it has to consider the characteristics of management support systems. MSS-specific are especially the tasks to be supported, i.e. managing and structuring.

A main basis for the structure and organization of each architecture model yields the enterprise architecture; which viewpoints and aspects have to be focused always depends on the specificity of the architecture model. The Semantic Object Model's (SOM) concept of the enterprise structure (e.g. see Ferstl and Sinz 1995, p. 2) is the basis for the specific architecture model. The development of an MSS – focusing management relevant viewpoints of the enterprise – is structured as depicted in Fig. 2-4 top.

The business concept – a model of a business system from an exterior viewpoint – is the basis for specifying the model-system of business processes, i.e. a corresponding model from an interior viewpoint. The specification of business applications systems can be derived from that model-system: Applications systems are identified and separated within the model-system of business processes; their purpose is to automate some part of a business process.

A detailed analysis of this *architecture for the development of MSS* makes obvious that an MSS which is developed on the basis of this architecture model is limited to supporting the managing task. In the case of managing, the management needs information concerning business processes and the assigned task bearers and resources (see Fig. 2-4 top). The model-system of business processes specified during the development process is – related to the automated sub-system – reflected within the MSS. As a consequence, such an MSS enables the management to direct and control the enterprise only by managing business processes. This happens by modifying the quantities of influence which are specified in the data base of the model-system of business processes. As managing activities by the management have no consequences for the structure of the model-system of business processes, they correspond to modifications in the related data base.

In order to support the structuring task, such an architecture is necessary, too. But an analysis of the management's structuring activities points out that they initiate structural modifications within the entirely enterprise architecture, especially within the model-system of business processes. The management uses for defining and executing structuring activities its – more or less well defined – knowledge about the enterprise structures, their relationships and interactions. The focus is on those aspects which are relevant for structuring the enterprise. As a consequence, a complete and comprehensive inclusion of all principal and concrete enterprise structures and their relationships has to be considered in the architecture additionally. Thus, a management-specific, detailed meta model of the enterprise architecture – realized as a repository – has to be provided in the automated sub-system of the MSS. It serves for explicitly modeling and completing this knowledge; its corresponding realization in a repository allows a computer-supported usage of this knowledge. Thus, an adequate consideration and support of the structuring task requires within the architecture for the development of MSS an *architecture of an MSS* (see Fig. 2-4 bottom).

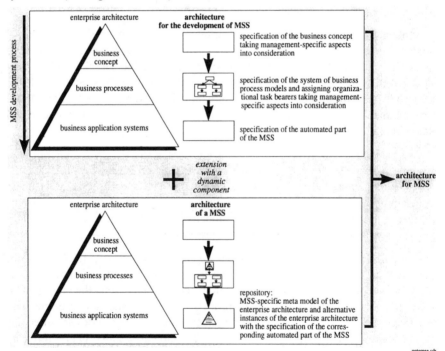

Fig. 2-4: Architecture Model for MSS.

The task-dependent extension of the architecture for the development of MSS with the architecture of an MSS meets the criterion „support both managerial task managing and structuring". The required integration of both architecture models in an architecture system – in the following, called *architecture for MSS* (see Fig. 2-4 right) – especially demonstrates the necessity of a specific architecture in order to increase the quality of MSS decisively.

The principal goal of an MSS – the entirely support of the management within its managerial tasks in order to retain the viability of the enterprise – determines the principal goals of an architecture for MSS, such as *management of complexity, optimization of an enterprise-wide management of resources, long-sighted retainment or reinforcement of competitive strength* and *winning certainty of management within the managerial tasks managing and structuring.* In addition, an architecture model for MSS has to support or even to initiate and force a *permanent business process re-engineering.* An *optimization of business processes* concerning consistency and freedom of redundancy as well as winning capabilities is another main target. Herewith, the goal *restructuring* or *reorganization* is strongly associated.

2.4 Synergies of Object-Orientation and Business Process-Orientation within the Architecture for MSS

Principally each of the paradigms, object-orientation as well as business process-orientation, offers essential benefits to MSS development and to a corresponding architecture model. Strengths of object-oriented approaches are for example:

- reusability: The possibility of reusing components – not only during implementation by means of class libraries – effects a more efficient system development because an MSS or at least a (model) component can be constructed using existing parts that need only little modification.
- flexibility: The principle encapsulation allows to encapsulate complex information in information objects and to make available – manager-specific – heterogeneous instances.
- explorability: An architecture model based upon object-oriented concepts allows using the same modeling structures during different levels of description. Within the level of requirements definition defined objects can be transformed into design objects and subsequently into software objects. Principally, conceptual changes can be realized due to the MSS without exception.
- extendibility: The use of the same modeling structures during different development phases in combination with inheritance and polymorphism has positive effects to extendibility.
- semantic richness: Object-oriented approaches fascinate through their high semantic richness and clearness – the result of encapsulating all essential information in objects. Different objects of business reality performing differing bu-

siness tasks can be classified by their business function, typed and interpreted as specialized objects. Object-specific information and tools can be defined as attributes and operators. This enables the association manager $\stackrel{\wedge}{=}$ object.
- user-friendliness: The use of existing models or systems or at least parts of them enable arguments by analogy. The manager is able to specify targets concerning the functionality and design of the MSS more easily and exactly.

A business process-oriented way of system development guarantees the support of operative work-flows by corresponding business application systems. Basing the Semantic Object Model (SOM), the specification of a business process consists of the following elements (see Fig. 2-5):
- The (transaction) services[1] the business process deals with have to be specified.
- A business object and a set of business transactions producing and transmitting the service have to be specified. Concerning the deliminaton of the world of discourse, internal and external objects are distinguished.
- Each transaction is associated with exactly two tasks. These tasks are assigned to the business objects which are connected by this transaction. Tasks can be classified into tasks of transformation or decision-making.
- Additionally, two kind of events are used to control the execution of tasks. Internal events connect tasks within an object; external events define environmental pre-conditions for the execution of tasks.

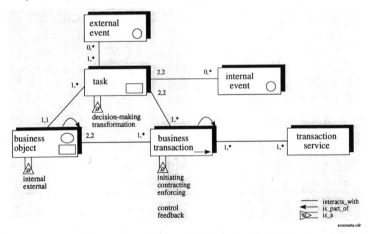

Fig. 2-5. Meta Model for Business Process Modeling in SOM (Ferstl and Sinz 1995, S. 216).

Amongst the principal advantages (e.g. see Ferstl and Sinz 1994b, S. 589f, Ferstl and Sinz 1995, S. 210f), there are special benefits for the architecture for MSS:

---

[1] In contrast to SOM, the meta object is named (transaction) service in order to avoid confusions concerning the term service.

- covering information needs: Integrating SOM allows the provision of all management-important viewpoints such as services, feedback controlling targets and
work-flows to an enterprise.
- provision of adequate tools for problem solving and decision making: The special structure of managerial business processes can be taken into consideration
and supported by appropriate tools.
- orientation towards value added and service: A both value chain- and service-
oriented viewpoint is the basis for different kinds of data analysis.
- reorganization and restructuring: The uncovering process of possibilities of ra-
tionalization that makes obvious capabilities of reorganization and restructuring
is supported.
- learning organization, especially concerning management structures: Adaptations towards information needs, management structures and styles that determine the basic managerial processes and – as a consequence – the positioning of
the enterprise are eased.
- controlling support: A transaction-oriented understanding of business processes
is the basis for an efficient application of newer controlling concepts such as
transactional cost analysis. These approaches supply the management with information allowing more distinctable processes of problem solving or decision
making.

The combination of both paradigms leads to synergies in general and especially
for this architecture:
- entirety: The possibility of defining object-specifically information and tools
depends on the feasibility of both the assignment of business process to specific
objects and the generation of management-specific viewpoints (,,business
process ownership").
- effectiveness/efficiency: Both structural and dynamic aspects – as a result of
business process-orientation – can be encapsulated in objects. Almost considering within the phase of conception, this fact allows an easy way of realizing
essential techniques for MSS such as news, data analysis, drill-down, exception
reporting or color-coding.

3 Meta Model of the Architecture for MSS

3.1 Survey of the Architecture for MSS

The meta model of the architecture – oriented towards the levels of SOM's enterprise architecture – primary consists of two levels (cg. Fig. 3-1):
- business concept: The identification of the enterprise's world of discourse, its
environment, its substantive goals and formal objectives, its success factors and

302

its value chain result in the business concept (Ferstl and Sinz 1994b, p. 6). Corresponding principal, management-specific detailed structures are described at this level.
- requirements definition: The provision of the general structures of a requirements definition takes place at the level of requirements definition. This level is divided into two sub-levels:
  - *business processes*: From a behavioral viewpoint, an enterprise consists of a set of business processes (Ferstl and Sinz 1994b, p. 6). This level covers the meta model for describing business processes management-specifically.
  - *business applications systems*: The purpose of a business application system is to automate some part of the business system. Application systems are identified and separated within the set of business processes (Ferstl and Sinz 1994b, p. 6). This level includes an object-oriented meta model for specifying MSS.

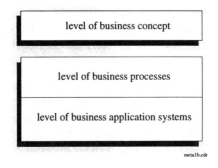

Fig. 3-1: Levels of the Architecture for MSS.

3.2 Level of Business Concept

An essential requirement to the level of business concept is to provide semantic rich, practical concepts supporting both managerial tasks managing and structuring in order to achieve the goals defined above. From an internal viewpoint, the business concept – documented to external as an unit – consists of several basic concepts. Specializations and details are added for winning informative value and semantic richness. In the following, basic concepts and related specialisations are derived successively.

The *value chain* serves as an instrument for enterprise analyses (Porter 1992, p. 64). Its development requires an investigation of all *value added activities*. Concerning their reference to value added, the activities can be classified into *primary* and *secondary* activities (Porter 1992, p. 65).

A *strategic business unit* (*SBU*) (Kreikebaum 1993, p. 113f, Wöhe 1990, p. 142f) disposes of its own value chain because it is directed towards a self-reliant market segment. Thus, value chains of different SBU are independent. An SBU can consist of either sub-SBU or activity units that are not equipped with the characteristics of an SBU – self-reliance of the market segment, independence of products or homogeneity of competition. Value added activities can be associated with exactly one activity unit because they always represent a piece cut out of an SBU.

The *goals* (Heinrich and Burgholzer 1990, p. 13, Martin 1990, p. 89f, Kargl 1993, p. 77) an SBU has to proceed have to be specified. In the main, the definition of a goal depends on an SBU. Although there might exist goals which are not directly associated with an SBU and are valid independent of the SBU's definition, those goals have to be considered in the scope of the definition of the SBU's goals. Goals can be classified either concerning their reference in *substantive goals* and *formal objectives* or relating their time-limit in *strategic*, *tactical* and *operative goals*. In addition, goals can be structured in a goal hierarchy.

Taking certain premises or possible restrictions into consideration, *strategies* (Tregoe and Zimmermann 1980, p. 17, Staehle 1991, p. 563) point out approaches how to realize some goals of an enterprise. Strategies can only be concreted or operationalized if they are based on a goal. Thus, the definition of strategies depends on certain goals. Like goals, strategies can be ordered hierarchically; lowest level strategies are called *basic strategies*. (*Global*) *enterprise strategies*, *business strategies* or *functional area strategies* describe specialisations of the meta object strategy.

304

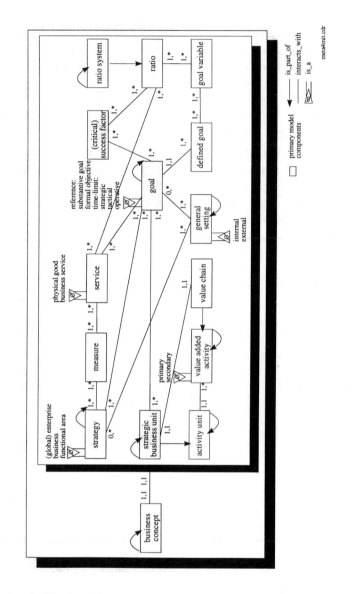

Fig. 3-2: Level of Business Concept.

Measures are used for operationalizing strategies; strategies and measures are directly connected. The result of some measures is expressed in some *services* – either a *physical good* or a *business service*. A service can only be produced by the execution of some measures. The specification of *general settings*, i.e. *internal* or *external* restrictions, effects the definition of goals decisively. The intensity of a restriction's influence is goal-specific. Goals without general settings do not exist; a contrary assumption is a contradiction to the intention of a goal. General settings serve especially the adjustment of resources, competitives, customers and suppliers. Short-termed general settings have greater effects to operative or tactical goals and functional area strategies; with respect to time constant or long-termed general settings determine both strategic goals and (global) enterprise strategies.

In addition to general settings, *critical success factors* (Heinrich and Burgholzer 1990, p. 234, Österle, Brenner and Hilbers 1992, p.374, Rockart 1980, p. 49ff) determine the definition of goals and strategies. Analyzing and identifying critical success factors especially allow the detection of deficit-ridden enterprise units. *Ratios* (Heinrich and Burgholzer 1990, p. 236, Rockart 1980, p. 49ff, Reichmann and Lachnit 1976, S. 705) serve precise, concentrated information about essential facts and trends within the enterprise and have to be defined for its critical success factors. An intended, systematic and structured summary of all ratios leads to a *ratio system* (Reichmann 1990, S. 19).

A comprehensive definition of goals requires the specification of a corresponding *defined goal* (Ferstl and Sinz 1992, p. 16); it describes the corresponding desired extension and the timely reference of the goal performance. Quantifiable *goal variables* (Ferstl and Sinz 1992, p. 16) allow measuring the degree of goal performance; another intention is to uncover the success critical aspects within the enterprise, i.e. the critical success factors. Ratios are also directly associated with services because they reflect quantifiable facts.

The entirely meta model for the level of business concept including all basic structures, specialisations and relationships is depicted in Fig. 3-2. Self-related arrows denote *is_part-of*-relations. Grey hatched meta objects describe primary model components, i.e. constructs without any pre-requirements concerning the structure of the meta model. It makes sense to start modeling by using these model components.

3.3 Level of Requirements Definition

The level of requirements definition is based on SOM that provides for both modeling business process and specifying business application systems meta models. For a detailed discussion see for example Ferstl and Sinz 1994b. The meta model for business process modeling is depicted in Fig. 2-5.

A business application system is specified by a schema of *related classes*. Each class is described by some *attributes* and *operators* (see Fig. 3-3).

306

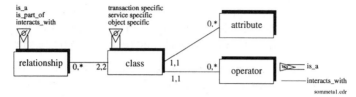

Fig. 3-3: Meta Model for Business Applications Systems Modeling in SOM (see e.g. Ferstl and Sinz 1994c, p. 15).

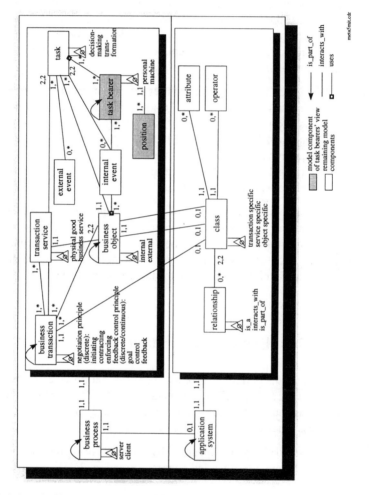

Fig. 3-4: Level of Requirements Definition.

In analogy to the level of business concept of the architecture for MSS, the essential requirement to the level of requirements definition is to provide semantic rich concepts allowing a comprehensive modeling. In order to guarantee this requirement, the management-important viewpoints have to be considered within the level of business processes. That means the SOM approach has to be adapted. There are only changes in the meta model for business process modeling because the management-specific tasks managing and structuring have to be considered at this task-dependent level:

- One adaptation is the extension to the concept *goal transaction* mentioned above.
- Another adaptation is the provision of the concepts *task bearer* and *position*. First, these extensions allow not only an optimization of business processes concerning the enterprise's structuring in operations, but moreover related to their organization structure. Then, all task bearers assigned to one position can be detected which eases the derivation of all position-oriented ESS. The entirety of all ESS describes the MSS.

The sub-levels of requirements definition can be connected from an both external and internal viewpoint:

- The tasks' degree of automation within a business process determines its support by a business application system.
- The internal relations are defined by SOM. *Services, transactions* and *objects* will lead to *service-specific, transaction-specific* and *object-specific classes* if the corresponding tasks are (partly) automated.

The meta model of the level of requirements definition including all basic structures, specialisations and relationships is visualized in Fig. 3-4.

3.4 Comprehensive View

In order to construct a comprehensive architecture model for MSS, the levels have to be associated by inter-relationships. Again, the levels of business concept and requirements definition can be connected from both an external and internal viewpoint:

- A *business concept* is the basis of derivation for several *business processes*. A business process is associated with exactly one business concept because modifications in a business concept mostly result in changed structures within the business process model.
- Essential ideas concerning the internal inter-relationships are given by the definition of the meta objects (see Fig. 3-5):

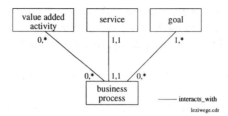

Fig. 3-5: Inter-relationships among Service, Goal, Value Added Activity and Business Process.

- A business process serves the production and transmission of specified services. Thus, the *service* specified within the business concept corresponds with a *business process*.
- The specification and execution of *business processes* always happens *goal*-directed. Conformity to the goals defined in the business concept is a substantive criterion.
- The execution of *value added activities* requires corresponding *business processes*. The possibility that a business process is not associated with value added activity and vice versa exists theoretically.

Fig. 3-6 illustrates the comprehensive meta model of the architecture for MSS.

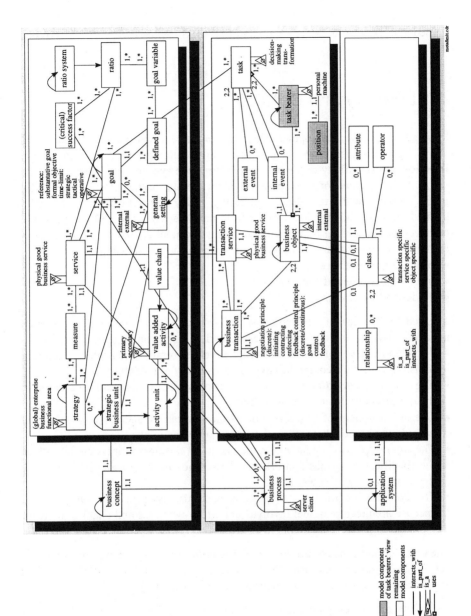

Fig. 3-6: Comprehensive View.

On the one hand, this model covers all aspects which have to be considered during the development of an MSS. On the other hand, this meta model is the basis for a repository realized in the MSS in order to support the management within its managerial task structuring. It makes sense to extend and detail this meta model. As a result, relationships that are only implicitly known or considered in this meta model, but contain additionally structuring capabilities become obviously. An extended and detailed architecture model of an MSS is presented in (Mehler-Bicher 1996).

4 Summary and Outlook

Principally system development requires models which combine economical knowledge with know-how about information technology in an optimal manner. Intention and specificity of the business information systems which have to be developed are decisive for constructing appropriate approaches. The example MSS points clearly out the task-dependent shifts. Today, the development of MSS is difficult process; existing MSS suffer from many lacks. MSS are too complex to be developed according traditional system development approaches. The realization that the quality of business application systems can be increased by

- basing an architecture model,
- using object-oriented concepts and
- modeling business process.

This led to the idea of developing an MSS-specific architecture. The presented meta model of the architecture for MSS considers management-specific aspects:

- The detailed level of business concept covers several – practically – important tools. They allow a comprehensive, objective way of modeling.
- The derivation of an initial system of business process models is supported by the definition of the inter-relationships.
- The initial system of business process models can be successively completed considering especially management-important viewpoints.
- The automated sub-system within the system of business process models allows the specification of position-oriented ESS. The entirety of all ESS represents the (automated part of the) MSS.
- The combination of object-orientation and business process-orientation in a specific architecture model involves important synergies and enables the development of better, more qualified systems.

The completion of the architecture model requires a corresponding procedure model. It can be shown that the SOM V-model (see Ferstl and Sinz 1995, p. 213) is applicable if MSS-specific extensions happen (see Fig. 0-7).

Architecture-specific production rules have to be defined in order to retain and guarantee consistency as well as integrity within the model-system. A detailed discussion of the production rules and the applicability of the SOM V-Model is given in Mehler-Bicher 1996.

In order to point out the architecture's applicability, an exemplary modeling has to take place. Besides the general applicability, one main result will be the deduction of adequate notation techniques for the business concept.

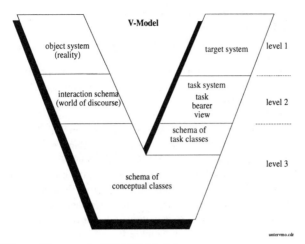

Fig. 4-7: MSS-specific, extended SOM V-Model.

References

Bleicher, K. (1991) *Das Konzept Integriertes Management*, Campus Verlag, Frankfurt, New York

ESPRIT Consortium AMICE (eds.) (1991) *Open Systems Architecture for CIM*, 2. Edition, Springer Verlag, Berlin et al.

Ferstl, O.K. and Mannmeusel, T. (1995) *Gestaltung industrieller Geschäftsprozesse*, in: Wirtschaftsinformatik 5/95, Vol. 37, p. 446 – 458.

Ferstl, O.K. and Sinz, E.J.: *Glossar zum Begriffsystem des Semantischen Objektmodells (SOM)* (1992) Bamberger Beiträge zur Wirtschaftsinformatik No. 11, Otto Friedrich-Universität Bamberg.

Ferstl, O.K. and Sinz, E.J. (1994a) *Grundlagen der Wirtschaftsinformatik*, 2. Edition, Oldenbourg, München, Wien.

Ferstl, O.K. and Sinz, E.J. (1994b) *From Business Process Modeling to the Specification of Distributed Business Application Systems – An Object-Oriented Approach*, Bamberger Beiträge zur Wirtschaftsinformatik No. 20, Otto Friedrich-Universität Bamberg.

Ferstl, O.K. and Sinz, E.J. (1994c) *Der Ansatz des Semantischen Objektmodells (SOM) zur Modellierung von Geschäftsprozessen*, Bamberger Beiträge zur Wirtschaftsinformatik No. 21, Otto Friedrich-Universität Bamberg.

Ferstl, O.K. and Sinz, E.J. (1995) *Der Ansatz des Semantischen Objektmodells (SOM) zur Modellierung von Geschäftsprozessen*, in: Wirtschaftsinformatik 3/95, Vol. 37, p. 209 – 220.

Heinrich, L.J. and Burgholzer, P. (1990) *Informationsmanagement: Planung, Überwachung und Steuerung der Informationsinfrastruktur*, 3. Edition, Oldenbourg, München, Wien.

Hildebrandt, K. (1992) *Ein Referenzmodell für Informationssystem-Architekturen*, in: Information Management 3/92, S. 6 – 12.

Kargl, H. (1993) *Controlling im DV-Bereich*, Oldenbourg, München, Wien.

Krcmar, H. (1990) *Bedeutung und Ziele von Informationssystem-Architekturen*, in: Wirtschaftsinformatik 5/90, S. 395 – 402.

Kreikebaum, H. (1993) *Strategische Unternehmensplanung*, 5. Edition, Kohlhammer, Stuttgart, Mainz, Köln.

Malik, F. (1992) *Strategie des Managements komplexer Systeme*, Haupt, Bern, Stuttgart.

Martin, J. (1990) *Information Engineering: Design and Construction*, Vol. 3, Prentice Hall, Englewood Cliffs, New Jersey.

Mehler-Bicher, A. (to be published in 1996) *Ein objekt- und geschäftsprozeßorientiertes Architekturmodell für Management Support Systeme*, Dissertation, Fakultät für Sozial- und Wirtschaftswissenschaften, Otto Friedrich-Universität Bamberg.

Österle, H., Brenner, W. and Hilbers, K. (1992): *Unternehmensführung und Informationssystem: Grundlagen der Unternehmensmodellierung*, 2. Edition, B.G.Teubner, Stuttgart.

Olle, T.W. et al. (1989) *Information System Methodologies: A Framework for Understanding*, Addison-Wesley Publishing Company, Reading, Massachusetts.

Porter, M.E. (1992) *Wettbewerbsvorteile (Competitive Advantage): Spitzenleistungen erreichen und behaupten*, 3. Edition, Campus, Frankfurt.

Reichmann, T. (1990) *Controlling mit Kennzahlen*, 2. Edition, Verlag Vahlen, München.

Reichmann, T. and Lachnit, L. (1976) *Planung, Steuerung und Kontrolle mit Hilfe von Kennzahlen*, in: Zeitschrift für betriebswirtschaftliche Forschung (zfbf), Vol. 28. (1976), p. 705 – 723.

Rockart, J.F. (1980) *Topmanager sollten ihren Datenbedarf selbst definieren*, in: Harvard Manager II/80, p. 45 – 58.

Scheer, A.W. (1992) *Architektur integrierter Informationssysteme: Grundlagen der Unternehmensmodellierung*, Springer, 2. Edition.

Scott-Morton, M.S. (1983) State of the Art of Resaerach in Management Support Systems, Working Paper CISR WP #107 and SLOAN WP #1473-83, Center for Information Systems Resaerch, Sloan School of Management, Massachusetts, Institute of Technology.

Staehle, W.H. (1991) *Management: Eine verhaltenswissenschaftliche Perspektive*, 6. Edition, Verlag Vahlen, München.

Sowa, J.F. and Zachman, J.A. (1992) *Extending and Formalizing the Framework for Information Systems Architecture*, in: IBM Systems Journal, Vol. 31, No. 3, p. 590 – 616.

Tregoe, B.B. and Zimmermann, J.W. (1980) *Top Management Strategy: What It Is and How to Make It Work*, Simon and Schuster, New York.

Ulrich, H. (1984) *Management*, Haupt, Bern, Stuttgart.

Wöhe, G. (1990) *Einführung in die Allgemeine Betriebswirtschaftslehre*, 17. Edition, Verlag Vahlen, München.

Zachman, J.A. (1987) *A Framework for Information Systems Architecture*, in: IBM Systems Journal, Vol. 26, No. 3, p. 250 – 263.

It's Time to Engineer Re-engineering: Investigating the Potential of Simulation Modelling for Business Process Redesign

George M. Giaglis and Ray J. Paul

Department of Computer Science and Information Systems, Brunel University, Uxbridge UB8 3PH, Middlesex, UK, gmg@aueb.gr, Ray.Paul@brunel.ac.uk

Abstract. While a multitude of methodologies have been proposed for carrying out BPR analyses, existing literature does not address the practical implementation issues for BPR in great detail. We argue that the complexity of business systems, their dynamic behaviour and the need to quantitatively evaluate among alternative designs call for the application of formal engineering techniques in business modelling. We propose discrete-event, stochastic simulation modelling as a suitable technique for that purpose and introduce a framework for integrating simulation in BPR exercises. Finally, we discuss the issue of Computer Aided Business Process Re-engineering (CABPR) and specify a set of functional requirements for CABPR tools.

Keywords. Discrete Event Simulation, Business Process Re-engineering, Business Process Modelling

1 Introduction

There have been few management concepts that have attracted such attention from both practitioners and academics, as has the concept of Business Process Re-engineering. BPR, otherwise referred to as Business Process Redesign or Process Innovation, was firstly introduced by Hammer (1990), Davenport and Short (1990) and Venkatraman (1991). Later the concept was further popularised by Davenport (1993) and Hammer and Champy (1993) who defined BPR as 'the fundamental rethinking and radical redesign of business processes to achieve dramatic improvements in critical, contemporary measures of performance, such as cost, quality, service, and speed'.

This definition implies a new look to organisations based on the processes they perform rather than the functional units, divisions or departments they are divided into. Processes are defined as 'structured, measured set of activities designed to produce a specified output' (Davenport 1993). In a definition directly relating to Porter's (1985) notion of value chain, Hammer and Champy (1993) defined processes as actions that, when put together, produce a result of value to the customer. The shift from functional-based to process-based organisational analysis is said to constitute a major paradigm shift in business and management science and therefore BPR has been categorised as a business revolution by its early advocates. Although this categorisation has been criticised (Mumford 1995, Earl 1994, Davenport and Stoddard 1994), there has also been suggested that BPR is a valuable management concept, whether being a revolutionary approach or not.

According to its definition, BPR calls for radical changes in the way an organisation operates. Compared to other similar approaches that have been proposed for business change, especially the concepts of Continuous Improvement (Harrington 1991) and Total Quality Management (Oakland 1993), BPR is broader in scope (typical BPR initiatives are cross-functional or even inter-organisational, see for example Riggins and Mukhopadhyay 1994) and requires the commitment of the senior management to ensure that strategic opportunities are identified and excessive changes can be applied. On the other hand, it is also characterised by a higher level of risk, requires significantly longer time to achieve and cannot be applied as a continuous based philosophy. According to Sadler's (1995) categorisation of organisational change, BPR is a change of transformational nature as opposed to the incremental changes advocated by most antecedent approaches.

The increasing business and academic interest in BPR has resulted in a multitude of methodologies on how it can be applied to organisations (see, for example, Wastell et al 1994, Harrison and Pratt 1993). However, the existing literature has been criticised for not providing adequate advice as to how to actually model business processes and design changes (Hansen 1994) or implement these changes in existing business environments (Galliers 1994, Burke and Peppard 1993).

In this paper, we argue that there is indeed a need to support the practical implementation issues of BPR, especially the tasks of business process modelling and evaluation of alternative process designs. In the following sections we analyse the context of the business environment and the characteristics of business processes, and we propose the application of formal engineering principles in the analysis and design of processes. Afterwards, the potential of computer simulation modelling as an engineering tool is assessed and a simple framework is introduced to integrate simulation as part of a BPR exercise. Finally, the desirability and feasibility of developing integrated computer supported tools for carrying out BPR exercises is discussed.

2 Process-based Organisational Approach

In his classic analysis of economic systems more than two hundred years ago, Adam Smith introduced the idea of the 'division of labour' as a means of achieving increased productivity by industrial workers. Although the economic and business environment has significantly changed since the rise of the industrial revolution, Smith's suggestions are still reflected on the organisational structure of most modern firms. This has resulted in companies being organised around functional units (departments or divisions), each with a highly specialised set of responsibilities. This form of organisation has the following characteristics (Blacker 1995, Sadler 1995):

- Units become centres of expertise which build up considerable bodies of knowledge in their own subjects.
- Even the simplest business tasks tend to cross functional units and require the coordination and cooperation of different parts of the organisation.
- Organisations tend to structure themselves in lots of hierarchical levels since responsibilities and duties at each level of operation need to be clearly and unambiguously defined.

The process-based organisational approach, introduces an alternative way of looking into organisations. It advocates the re-unification of separate business tasks in a way that they constitute a set of activities with clearly added value to their (internal or external) customers.

Although one could argue that processes should be the obvious way of analysing organisations since they represent how things are actually done, this has not been the case, especially in large organisations. Almost every method of analysing and documenting businesses (e.g. organisational charts, policy manuals) is based on a functional approach which may well indicate levels of hierarchy and responsibilities, but does not provide any insight as to how certain tasks are performed. So, for example, the actual procedure of handling an invoice or fulfilling a customer order might be unknown or unclear to a senior manager of a large corporation, although it is probably a very important process for his firm.

Typical examples of major business processes include the purchasing of raw materials from suppliers, the use of these materials to produce goods and/or services, the delivering of these goods/services to customers, the acquisition of new customers, the development of new products according to customer needs, etc. It is obvious that each of these processes requires the cooperation and synchronisation of different functional units in order to be successfully performed. It is also typical for a business process to cross organisational boundaries and extend to third parties (customers, suppliers, etc.).

Furthermore, processes can be described in different levels of detail depending on the abstraction put into analysing the organisation. Typical business processes like those identified above, can be divided into sub-processes which can be further split until the level of individual business tasks. The levels of analysis depend on the organisation's size and type, but it has been reported that the major business processes (i.e. those in the higher level of abstraction) are typically similar (Blacker 1995) and even very large organisations can identify only a limited number of business processes in this stage (Davenport 1993).

3 Engineering a Process: The Need for a Formal Approach in Process Modelling

BPR calls for radical transformation of the way business is done, by adopting a novel view of organisations based on the processes they perform and then changing these processes to achieve maximum overall performance. What BPR does *not* say, is how to actually identify which processes must be changed, how to specify alternative process designs, and how to select between them in order to decide on which changes to implement. In other words, existing BPR methodologies do not tackle the actual problem of 're-engineering' the processes. Even Hammer and Champy (1993) in the last page of their best-selling book on BPR, admit that the issues of actual BPR implementation are probably the most problematic and their analysis 'goes beyond the scope of a single book'.

In strict terms, to engineer a business process (or anything else, for that matter) means to apply scientific principles in its analysis, design and implementation. Taking this into account, many researchers argue that business processes have never been engineered (Hammer 1990); rather, they have evolved over time (Hansen 1994). This can easily explain why they have never been standardised as well; each organisation develops its own proprietary procedures for accomplishing certain tasks when a need evolves. These procedures can be modified when internal or external factors change, but usually without applying any formal or scientific techniques for that purpose. The development, refinement and obliteration of business processes is a procedure built into the ethos and business culture of each organisation.

It has been argued (Willcocks and Smith 1995, Galliers 1993) that businesses and business processes are sufficiently complex systems and therefore carefully developed models are necessary for understanding their behaviour in order to be able to design new systems or improve the operation of existing ones. We can distinguish the basic requirements of the decision makers regarding the modelling process (Streng 1994, Wastell et al 1994) in two separate areas: 'Technical' requirements which refer to those needs that call for the application of engineering principles, and 'Political' requirements which refer to the needs that emerge from the social nature of business systems. Table 1 contains a, by no means exhaustive, list of these requirements.

Although the above requirements highlight the multidisciplinary, holistic view that is central to the notion of BPR (Glykas and Lytinas 1995), almost all BPR methodologies that have been proposed up to now utilise partial approaches in their effort to engineer business processes. Most of them adopt a behaviourist conceptual position and they concentrate on philosophical, people-oriented, or culture-oriented methods that emphasise increasing communications about processes. There is still no universal or proven methodology (Earl 1994) to incorporate both the technical requirements and the social dimension of business systems, the main problem being the inability to address the issues of formal process representation and quantitative measurement of the impact of proposed changes.

In view of the above, the application of scientific techniques in business process analysis and design is deemed necessary if processes are really to be 'engineered'. However, the application of scientific techniques must also take into account that businesses are not solely technical systems; they include human participation as well. Any attempt to radically modify business structure and mode of operation must be carried out with considerably more care and sensitivity if the results are to be satisfactory and acceptable and the transition is to be stable. In other words, a process structure that may appear to 'optimise' business performance theoretically (as a result of the engineering process) might not be applicable in practice due to factors, such as user rejection, political or legislative requirements, etc. Therefore, the need is clearly for techniques that can efficiently address both requirements. In the next section we will investigate the potential of simulation modelling for this purpose.

318

Technical requirements	Processes need to be formally modelled and documented
	Models should be easily updatable to follow changes in actual processes
	Models should be reusable and reconfigurable
	Modelling should take into account the stochastic nature of business processes, especially the way in which they are triggered by external factors
	There is a need to quantitatively evaluate the value of proposed alternatives
	The evaluation is highly dependent on the objectives of the particular study
Political requirements	Modelling and decision making should take into account such factors as legislation restrictions, user acceptance of changes, etc.
	Modelling should be flexible to allow for different interpretation of analysis results according to specified objectives
	There is a need to communicate alternatives to the senior management as well as to the end-users
	A holistic approach is necessary to identify implicit interdependencies among processes. On the other hand, parts of the organisation should be able to use partial models to assess their own performance
	Modelling tools should be easy to use to allow users of the processes to be involved in the modelling process

Table 1: Requirements of Process Modelling

4 Simulation Modelling and BPR

The idea that Information Technology (IT) enables the organisational rethinking that is advocated by BPR, is present (implicitly or explicitly) in the existing BPR literature. However, when it comes to the actual implementation of a BPR exercise, the potential of IT is usually ignored and most methodologies utilise other, non-computer supported, techniques for process modelling and evaluation. Such techniques include flow analysis (e.g. flowcharting or time analysis) and prototyping of processes.

There have been a few cases where novel approaches to BPR implementation have been introduced. Such examples include the use of object-oriented technology (Wang 1994), workflow software, etc. However, even these approaches refer to the automation of business processes *after* BPR implementation and not to the automation of the actual procedure of carrying out BPR. In the present section we will investigate the potential of computer simulation as a suitable technique for business modelling. We will start by a brief introduction to simulation and we will afterwards try to assess its potential as an engineering tool. After a brief presentation of Davenport's generic framework of BPR, we will modify this framework to include simulation as an integral part of a BPR project.

4.1 Simulation Modelling

Shannon (1975) has defined simulation as 'the process of designing a model of a real system and conducting experiments with this model for the purpose, either of understanding the behaviour of the system or of evaluating various strategies (within the limits imposed by a criterion or set of criteria) for the operation of the system'.

Practical simulation modelling will usually originate in a management perception of a problem requiring some decision or understanding (Paul and Balmer 1993). The problem may concern or involve the operation of some complex system on which direct experimentation may be impractical on grounds of cost, time or some human restriction. Simulation models provide a potentially powerful tool for conducting controlled experiments by systematically varying specific parameters and rerunning the model (Pidd 1992, Paul and Doukidis 1992).

The major advantages of simulation over other operational research techniques are described by Law and Kelton (1991) as follows:
- Most complex, real-world systems with stochastic elements cannot be accurately described by a mathematical model that can be evaluated analytically. Thus, simulation is often the only type of investigation possible.
- Simulation allows to estimate the performance of an existing system under some projected set of operating conditions.
- Alternative proposed system designs (or alternative operating policies for a single system) can be compared via simulation to see which best meets a specified requirement.
- In a simulation we can maintain much better control over experimental conditions than would generally be possible when experimenting with the system itself.
- Simulation allows us to study a system with a long time frame (e.g. an economic system) in compressed time.

- Simulation, especially when combined with graphical animation and interaction capabilities, facilitates better understanding of a system's behaviour, of the impact of proposed changes and allows for better communication of results.

However, simulation modelling of a complex system is not meant to provide 'the answer' to modelling problems; rather, it is (or should be) used to make inferences about the system of interest (Law and Kelton 1991). In other words, its major contribution is to structure the problem and provide the decision makers with the quantitative information they need in order to make informed decisions. This is critical to the success of decision making: Blattberg and Hoch (1990) found that the combination of forecasting models with managerial intuition results in a significant improvement in the quality of decisions. What is even more surprising, is that the above is true even when the models are of poor quality!

4.2 Simulation as an Engineering Tool

The very definition of simulation reveals its great potential as a tool for BPR. Indeed, simulation modelling of an organisation's processes can help towards understanding the behaviour of the existing system, identification of problematic tasks and also makes experimentation with alternative processes easier, directly comparable and less risky. Some characteristics of simulation that make it ideal for business process modelling, include:

- Simulation modelling techniques are by nature process-oriented. A process in simulation terminology is defined as a time-ordered sequence of interrelated events which describes the entire experience of an 'entity' as it flows through a 'system' (Law and Kelton 1991), a definition closely related to those of business processes presented above.
- Simulation allows for experimentation with any element of a business system. This means that the behaviour of both technical (e.g. machines, systems) and non-technical (e.g. men) components of a business can be incorporated in a simulation study.
- Simulation helps to define deficiencies early in the design process when correction is easily and inexpensively accomplished.
- Simulation models can be easily updated to follow changes in the actual system, thus enabling model maintenance and reusability.
- Simulation models can improve decision quality through their consistency and objectivity.
- Simulation models can help the decision makers generate and communicate ideas and interact with the model to immediately assess the impact of proposed changes.
- The stochastic nature of business processes (i.e. the 'random' way in which they are triggered by external events) can be modelled in a simulation study.

- The analysis of results can be targeted to match the objectives of specific studies.
- Simulation allows the decision maker to obtain a 'system-wide' view of the effects of 'local' changes in a system and allows for the identification of implicit dependencies between parts of the system.
- Finally, simulation encourages a cultural shift in the way modelling is perceived in an organisation, by means of continuous measurement and evaluation of business activities.

As a technique, simulation is one of the most widely used in operations research and management science (Harpell et al 1989, Forgionne 1983). There are simulation models that try to represent specific aspects of a business environment (e.g. finance or production) and they are usually quite focused. One such category are financial models which are mainly concerned with risk analysis (Seila and Banks 1990). Another category is manufacturing systems, where their complexity and dynamic behaviour are the main reasons for the use of simulation as a modelling tool for facilitating their design and assessing operating strategies (Hlupic 1994). There are also examples where there is a concentration on a specific business category like the SERTS system which deals exclusively with economic and financial simulation of small businesses (Cohran et al 1990).

However, there are also simulation models of a more generic nature that have been used in business process re-engineering efforts. An example is MOSES (Pruett and Vasudev 1990) which simulates all manufacturing organisation processes, categorised as marketing, production, inventory and accounting. It models these processes, simulates the basic relationships with each other, and shows the immediate effect of a multitude of decisions made with respect to those processes. MOSES provides the opportunity for the manager in charge of one process (e.g. inventory) to see the impact his policies might have on the organisation's other major processes (e.g. production).

Another integrated simulation environment is the PIT framework (Streng 1994) that assumes that every value-added system consists of three components: the basic processes, the infrastructure of the system and the triggers that cause the processes to happen. A simple simulation model has been developed to assess re-engineering alternatives in a fictional firm. The system operates with advanced visual interactive capabilities that enable the generation of various alternatives for dealing with BPR during management brainstorming sessions.

Simulation has also been identified as a suitable technique for Continuous Improvement projects. MacArthur et al (1994) have showed that once in use, simulation models encourage a culture of measurement that supports continuous process improvement. The authors endorse an approach of: (a) gathering and analysing data on the elements and inter-relationships in the value chain and then (b) producing a simulation model to measure the performance of the business key functions.

The same basic idea is followed by Nissen (1994) who presents a small scale simulation of business processes for selecting among alternative IT investments. This approach emphasises modelling of and experimentation with alternative organisational processes for purposes of redesign and re-engineering. Prospective Information Systems are then designed to suit the chosen process schema.

Finally, simulation modelling has been used (Mylonopoulos et al 1995a, 1995b) to assess the expected benefits of inter-organisational changes made possible by the use of Information Technology. The authors have developed a model that simulates trading between a number of companies along a value chain and used it as a vehicle for assessing efficiency gains introduced by the use of EDI in specific industry sectors. They argue that similar simulation systems can be developed for evaluating the expected impacts of organisational changes in general, by the introduction of reusable and reconfigurable simulation modules that can be incorporated in simulation libraries and be subsequently used in many similar studies.

From the above analysis, it is obvious that simulation modelling has already been identified as a suitable tool for business modelling and has been used successfully in individual BPR studies. However, a comprehensive BPR methodology that uses simulation modelling to evaluate alternative redesign scenarios and capture business process performance has not yet been developed (MacArthur et al 1994). In the following paragraphs, we will introduce a simple framework for integrating simulation in BPR implementation. This framework is based on Davenport's generic framework for BPR which is briefly presented next.

4.3 A Simple Framework for BPR

Davenport (1993), one of the originators of BPR, proposed a simple framework for BPR implementation consisting of five steps (Figure 1):
- A BPR project should start by identifying the processes that are performed by the organisation so as those that are more suitable for re-engineering can be selected.
- In order to design and perform the re-engineering process, senior management should identify the critical factors that must be taken into account before re-engineering, i.e. the primary enablers of BPR in the individual organisation. These can be the use of Information Technology, the human resources of the organisation, etc.
- BPR can only be meaningful if it improves businesses in a way that is consistent with their strategy (Fiedler et al 1995, Galliers 1993). Therefore, it is essential for every company that engages in a BPR project to have formulated a clear business strategy that will inspire a vision to all parts of the business - the processes. This vision will act as a guide both to the identification of problematic processes and the design of alternative ones. Therefore, developing

visions (both at the organisation and at the process levels) is the next step of a BPR exercise.

- The fourth step proposed is the understanding of the way existing processes are performed. A view of the existing processes is necessary in order for problems and pitfalls to be identified and not repeated. Moreover, especially in the case of complex and specialised business processes, the re-engineering team might need to get familiar with the nature of work before designing the new processes. Finally, modelling and understanding current processes will provide a basis for comparing the impact of new processes to specific metrics, such as cost reduction, time reduction or profit increase.

- According to the results of the previous steps, the re-engineering team should be able to produce a number of alternative processes that should be comparatively evaluated before deciding on a particular solution. Once the decision-making process is complete, the organisation needs to formulate a migration strategy towards the selected solution and start implementing the new organisational structures and systems that will support the new processes.

As we argued above, there is clearly a need to support the actual part of modelling the processes and carrying out the studies that will help the management in the decision-making process (steps 1, 4, and 5 of Davenport's framework). In the next paragraph, we will try to identify the ways in which computer simulation modelling could assist these tasks and integrate simulation in a more detailed framework of BPR.

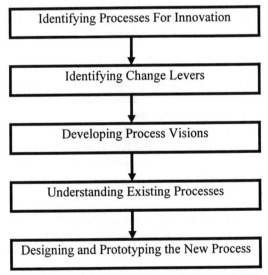

Fig. 1: A Simple Framework for BPR (Davenport 1993)

4.4 A Framework for Simulation and BPR

Simulation modelling can be a useful tool in every stage of a BPR project. In terms of the framework presented earlier on, we can identify the following potential applications of simulation:

Step 1: Identifying Processes For Innovation
- Simulation can be used for modelling either the major processes of an organisation or the detailed sub-processes according to modelling needs (hierarchical simulation modelling). This can help towards the identification of: (a) individual tasks with little or no business value, and (b) problems in major processes that occur due to the lack of alignment of their components.
- Simulation models are basically software models and can therefore be stored and easily updated to follow actual changes in business processes. In this way, the organisation can keep a repository of its process models that can be re-used in subsequent studies.

Step 2: Identifying Change Levers
- Simulation modelling views systems in a management perception and can therefore be used to experiment with all of the components of a business system, whether technical or not. For example, various BPR enablers (including Information Technology, organisational structure and human resource management) can be investigated by simulation experiments.
- In terms of IT, simulation can be used to experiment with different structures of Information Systems and evaluate the impact they might have on business operations. This can provide the quantitative information that is necessary for IT investment appraisal.
- In terms of organisational structure, simulation can be used to experiment with different layouts of the system (allocation of responsibilities, hierarchical structures, movement of processes to different departments, etc.) without bearing the cost of applying these changes in practice.
- In terms of human management, simulation can provide invaluable information on the optimum allocation of labour force to individual processes and on the necessary characteristics of this labour force, according to the resources available (or affordable) by the organisation.

Step 3: Developing Process Visions
- Simulation cannot be used as a tool in this step, but instead can make use of the results of it to identify the output variables that will provide the metrics against which business performance should be evaluated. For example, if cost reduction is the major specified business objective, then individual processes should be evaluated upon such characteristics as lead time, labour cost, management cost, etc.

- Furthermore, simulation encourages a culture of measurement in the organisation which can help the management employ a policy of continuously evaluating system performance and adapting business visions accordingly.

Step 4: Understanding Existing Processes
- Simulation modelling techniques are naturally process-oriented. Therefore, the techniques and tools that are used in simulation modelling (e.g. activity cycle diagrams, event graphs, etc.) can be directly applied to BPR modelling as well.
- Simulation modelling provides an easy-to-use mechanism for structuring and formally defining the processes performed in an organisation (AS-IS model), a vital task for the beginning of a successful BPR project.
- Simulation modelling can then be used to understand the behaviour of the existing system and to identify pitfalls, bottlenecks and other problematic situations that need to be addressed. By identifying problems in an early stage, the direction for change becomes more apparent and changes can be applied with less cost.
- In this sense, simulation of existing processes and analysis of the results can also be used as a vehicle for initiating short-period process improvement initiatives, especially to processes where complete re-engineering is not deemed necessary.
- Furthermore, by simulating the existing structure and work of an organisation, one can obtain useful quantitative information that can be used not only as a basis for comparing alternative solutions, but also as a means of evaluating the validity of the AS-IS model itself (i.e. the degree to which the model corresponds to the real system it is supposed to represent).

Step 5: Designing and Prototyping the New Process
- Simulation can be used to model alternative solutions, to conduct controlled experiments and obtain quantitative data that can be used by the management to decide upon the best re-engineering solution among those identified (SHOULD-BE model).
- The models of both the existing system and the proposed alternative designs can be used by decision makers during management brainstorming sessions. In this way (especially if the models incorporate graphical representation of processes, animation, and interaction characteristics) decision makers will be able to communicate their ideas and directly generate results on the expected impact of proposed changes.

Based on the above observations, a modified form of Davenport's framework to include simulation modelling as part of a re-engineering project, is illustrated in Figure 2. The greyed areas in the framework represent the steps of the project where simulation can provide valuable assistance to the re-engineering team and the decision makers.

326

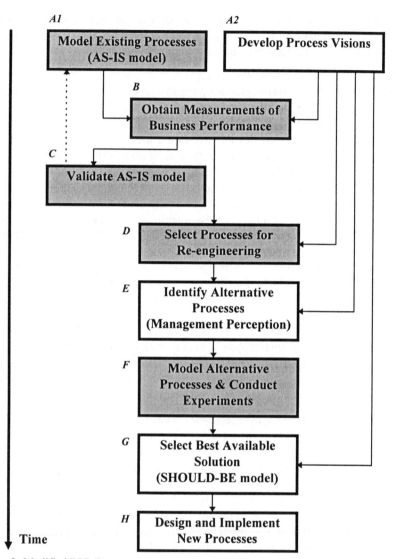

A1 Model Existing Processes (AS-IS model)

A2 Develop Process Visions

B Obtain Measurements of Business Performance

C Validate AS-IS model

D Select Processes for Re-engineering

E Identify Alternative Processes (Management Perception)

F Model Alternative Processes & Conduct Experiments

G Select Best Available Solution (SHOULD-BE model)

H Design and Implement New Processes

Time

Fig. 2: Modified BPR Framework

The proposed framework not only details Davenport's one, but also suggests a different order of implementing the steps required for a BPR project when simulation is used as a modelling and experimentation tool. Indeed, we suggest that the re-engineering project should start by a combination of modelling the existing organisation structure (step A1) and developing a clearly identified set of business and process visions (step A2). The former activity relates with simulating the existing processes, while the latter is a management task which is closely related to the business culture and objectives.

The combination of these tasks allows for the specification of the simulation output analysis that should be conducted in the AS-IS model (step B). This is very critical to the success of the whole study, as simulation models can generally be used to obtain a variety of quantitative information on system performance. However, what is needed in a BPR study is a carefully selected subset of information that is closely related to the business objectives. Therefore, the results that should be obtained by running the AS-IS model should be aligned with the visions of the management.

Afterwards, the developed AS-IS model should be validated, i.e. compared to the real world system whose behaviour the model is supposed to imitate (step C). The validation of the model might indicate errors in modelling, in which case the project should be halted and a new model of the existing system be developed and re-validated.

On the other hand, if (and when) the simulation model proves to be valid and acceptable by the management, the project can proceed by identifying processes that are suitable for re-engineering (step D). The selection of these processes should be based on the results obtained by simulating the AS-IS scenario (priority should be given to more problematic processes), the business vision (which highlights the strategic importance of each individual process) and, of course, the resources available by the organisation.

Based on the selection of problematic processes and the strategic visions of the organisation, an identification of alternative processes (possible solutions) should be the next step (step E). In this stage, this will include only the management's perception of the possible solutions (i.e. we are not looking at representing the solutions into simulation models at this stage). This will help towards identification of alternatives that will be acceptable both by the management and the end-users and, at the same time, do not contradict with requirements of existing legislation or other restrictions. These alternatives should be matched with those derived by applying statistical experimental design techniques to the simulation model. The combination should provide a final set of alternative process designs that will be both scientifically valid and 'socially' acceptable.

The next stage of the BPR project should be to model the solutions and conduct experiments with alternative process scenarios (step F). Ideally, the results from those experiments should help the decision makers in identifying an optimal solution, according to the strategic visions of the organisation (step G). The BPR project ends by designing a migration strategy towards the selected solution as well as a detailed implementation plan (step H).

5 Discussion and Future Research: Developing Computer-Aided BPR Tools

BPR calls for radical organisational change by means of (a) focusing analysis to business processes rather than functions and (b) re-engineering these processes to maximise performance.

The process-based organisational approach seems to be useful as it allows for the identification of problematic situations and/or new business opportunities. However, the approach of 're-engineering' the processes might just not be feasible. Technically speaking, one cannot re-engineer something that has never been engineered in the first place. So, if organisations really want to be successful in redesigning the processes they perform, they must put adequate effort in the actual modelling and analysis of their operations.

We have argued that the inherent characteristics of businesses (complex systems of dynamic behaviour, but at the same time with a strong social content) call for the application of modelling techniques that combine:
a) scientific (engineering) principles which facilitate the quantitative measurement of various indexes of business performance
b) management perception of systems to allow consideration of social and other non-quantifiable requirements.

We have also argued that stochastic simulation modelling seems to be the operational research technique which best qualifies for the purpose of robust modelling of business processes. Simulation can also be used for experimentation purposes and help decision makers during the analysis procedure. However, simulation cannot handle all parts of a BPR exercise. For example, typical simulation modelling does not include the actual analysis of process performance (this has deliberately been left to decision makers, as stated earlier on) nor does it provide a means of linking the modelling process with the 'real-world' input data required to produce valid models of businesses.

Although simulation can be a first step towards automating the procedure of carrying out a BPR analysis, we believe that there exists a great potential for the integration of it with other computer-supported tools to fully automate all steps required in the modelling and analysis process. Such tools, henceforth referred to as Computer Aided BPR tools (CABPR) should incorporate the following procedures inherent in BPR analyses:
- capturing the data from the organisation to facilitate design of processes
- modelling the existing processes and experimenting with alternative designs (simulation is the primary tool here)
- analysing the output data to automate the procedure of selecting optimal design configurations
- assisting the end-users in actually carrying out the processes designed as part of BPR

- facilitating continuous monitoring of process performance to identify opportunities for further improvements.

Figure 3 illustrates a typical configuration of CABPR tools.

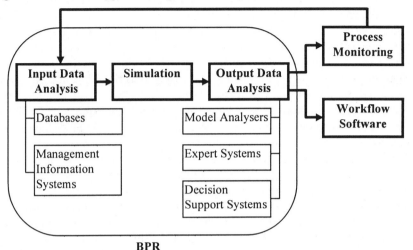

Fig. 3: Configuration of a CABPR tool

Of course, the integration of the above tasks is far from trivial, as some of them require significant further research as to how they could be effectively accomplished. For example, automating the procedure of output analysis of simulation-generated data is still a major research question in applied simulation modelling.

In any case, CABPR tools must also have some basic functional characteristics identified earlier on as major requirements of the BPR process. These characteristics include:

- *Graphical representation of processes.* There should be at least three distinct functional elements of processes, namely entities (like people, machines, etc.), activities (that entities are involved in) and queues (in which entities are placed when they are idle or when they are waiting for system resources to become available). CABPR tools should allow for the graphical representation of the processes by means of combining system elements in a distinctive, user-friendly way.
- *Built-in Libraries.* CABPR tools should have extensive built-in libraries of icons to represent the workings of typical business blocks (i.e. typical kinds of entities, activities, queues, or even whole processes).
- *Iconic Customisation of building blocks.* The graphical representation of different business system attributes should be fully customised to allow users to create their own icons to represent features not included in the built-in libraries.
- *Animation.* The main reason for the necessity to incorporate animation characteristics in CABPR tools is its ability to communicate the essence of a

simulation model to managers and end-users, thus increasing the model's credibility.

- *Interactive Simulations.* Users of CABPR tools should also be able to interact with them when simulation scenarios are run, in order to view the effects of proposed changes during the execution time.
- *Hierarchical Decomposition of Models.* The model should allow for performance assessment of individual processes as well as integration of the results to assess global performance.
- *Reusability.* The models developed with CABPR tools should be designed so as to be easily updated when processes change and then reused in subsequent studies, without the need to model the whole organisation from scratch. Object Orientation techniques might be appropriate to achieve high reusability of models.
- *Input Data Analysis.* This refers to the tool's ability to interact with external sources of business data (like databases and business applications) from which to obtain quantitative information to be converted into appropriate statistical distributions which will be used in simulating the behaviour of the business.
- *Output Data Analysis.* CABPR tools should also be linked to other applications which could make use of the data produced by simulating the business behaviour. This refers mainly to analysis and decision tools (like Expert Systems and Decision Support Systems) or to tools that will support carrying out the processes (like Workflow Software).
- *Process Documentation and Reporting.* CABPR tools must be capable of providing documentation of models by means of automatically generating reports on model structure.
- *Ease of Use.* Finally, building up models and analysing processes should be made accessible to non-experts and should therefore be made a relatively simple procedure.

References

Blacker, K. (1995) *The basics of Business Process Re-engineering*, Edistone Books, Birmingham.

Blattberg, R. C. and Hoch, S. J. (1990) Database Models and Managerial Intuition: 50% Model + 50% Manager, *Management Science*, vol.36, no.8, pp.887-899.

Burke, G. and Peppard, J. (1993) Business Process Redesign: Research Directions, *Business Change and Re-engineering*, vol.1, no.1, pp.43-47.

Cohran, M., Richardson, J. and Nixon, C. (1990) Economic and financial simulation for small business: a discussion of the small business economic, risk, and tax simulator, *Simulation*, vol.54, no.4, pp.177-188.

Davenport, T. H. and Stoddard, D. B. (1994) Reengineering: Business Change of Mythic Proportions?, *MIS Quarterly*, vol.18, no.2, pp.121-127.

Davenport, T. H. (1993) *Process Innovation: Reengineering Work Through Information Technology*, Harvard Business School Press, Boston, Massachusetts.

Davenport, T. H. and Short, J. E. (1990) The new industrial engineering: information technology and business process redesign, *Sloan Management Review*, Summer, pp.11-27.

Earl, M. J. (1994) The new and the old of business process redesign, *Journal of Strategic Information Systems*, vol.3, no.1, pp.5-22.

Fiedler, K. D., Grover, V. and Teng, J. T. C. (1995) An Empirical Study of Information Technology Enabled Business Process Redesign and Corporate Competitive Strategy, *European Journal of Information Systems*, vol.4, pp.17-30.

Forgionne, G. A. (1983) Corporate Management Science Activities: An Update, *Interfaces*, vol.13, no.3, pp.20-23.

Galliers, R. D. (1994) Information Systems, Operational Research and Business Reengineering, *International Transactions of Operational Research*, vol.1, no.2, pp.1-9.

Galliers, R. D. (1993) Towards a Flexible Information Architecture: Integrating Business Strategies, Information Systems Strategies and Business Process Redesign, *Journal of Information Systems*, vol.3, no.3, pp.199-213.

Glykas, M. and Lytinas, N. (1995) From Rapid Growth to Maturity: The Need for Holistic Methodologies in Business Process Redesign, In *The Proceedings of the 3rd European Conference on Information Systems*, Athens, Greece, pp.525-545.

Hammer, M. and Champy, J. (1993) *Reengineering the corporation: a manifesto for business revolution*, N. Brealey Publishing, London.

Hammer, M. (1990) Re-engineering work: don't automate - obliterate, *Harvard Business Review*, July/August, pp.104-112.

Hansen, G. A. (1994) *Automating Business Process Reengineering: Breaking the TQM Barrier*, Prentice-Hall, Englewood Cliffs, New Jersey.

Harpell, J. L., Lane, M. S. and Mansour, A. H. (1989) Operations Research in Practice: A Longitudinal Study, *Interfaces*, vol.19, no.3, pp.65-74.

Harrington, H.J. (1991) *Business Process Improvement: The Breakthrough Strategy for Total Quality, Productivity and Effectiveness*, McGraw-Hill, New York.

Harrison, B. D. and Pratt, M. D. (1993) A Methodology for Reengineering Businesses, *Planning Review*, vol.21, no.2, pp.6-11.

Hlupic, V. (1994) A comparative analysis of six manufacturing simulators, In *The Proceedings of The 4th International Conference on Information Systems Development*, Bled, Slovenia, pp.693-696.

Law, A.M. and Kelton, D.W. (1991) *Simulation Modelling and Analysis*, 2nd ed., McGraw-Hill, New York.

MacArthur, P. J., Crosslin, R. L. and Warren, J.R. (1994) A strategy for evaluating alternative information system designs for business process reengineering, *International Journal of Information Management*, vol.14, no.4, pp.237-251.

Mumford, E. (1995) Book Review of Reengineering the Corporation: A Manifesto for Business Revolution (Hammer, M. and Champy, J.), *European Journal of Information Systems*, vol.4, no.2, pp.116-177.

Mylonopoulos, N. A., Doukidis, G. I. and Giaglis, G. M. (1995a) Assessing the expected benefits of Electronic Data Interchange through Simulation Modelling Techniques, In *The Proceedings of the 3rd European Conference on Information Systems*, Athens, Greece, pp.931-943.

Mylonopoulos, N. A., Doukidis, G. I. and Giaglis, G. M. (1995b) Information Systems Investments Evaluation through Simulation: The Case of EDI, In *The Proceedings of the*

8th International Conference on EDI and Interorganisational Systems, Bled, Slovenia, pp.12-26.

Nissen, M. E. (1994) Valuing IT through virtual process measurement, *In The Proceedings of the 15th International Conference on Information Systems*, Vancouver, Canada, pp.309-323.

Oakland, J. S. (1993) *Total Quality Management: The route to improving performance*, 2nd ed., Nichols Publishing, New Jersey.

Paul, R. J. and Balmer, D. W. (1993) *Simulation Modelling*, Chartwell-Bratt Publishing, Lund, Sweden.

Paul, R. J. and Doukidis, G. I. (1992) Artificial Intelligence and Expert Systems in Simulation Modelling. In *Artificial Intelligence in Operational Research* (Doukidis, G. I. and Paul, R. J., eds.), Macmillan, pp.229-238.

Pidd, M. (1992) *Computer simulation in management science*, 3rd ed., John Wiley.

Pruett, J. and Vasudev, V. (1990) MOSES: manufacturing organisation simulation and evaluation system, *Simulation*, vol.54, no.1, pp.37-45.

Porter, M. (1985) *Competitive Advantage: Creating and Sustaining Superior Performance*, Free Press, New York.

Riggins, F. J. and Mukhopadhyay, T. (1994) Interdependent Benefits from Interorganisational Systems: Opportunities for Business Partner Reengineering, *Journal of Management Information Systems*, vol.11, no.2, pp.37-57.

Sadler, P. (1995) *Managing Change*, Kogan Page, London.

Seila, A. and Banks, J. (1990) Spreadsheet risk analysis using simulation, *Simulation*, vol.55, no.3, pp.163-170.

Shannon, R. E.(1975) *Systems Simulation: the art and the science*, Prentice-Hall New Jersey.

Streng, R. J. (1994) BPR needs BIR and BTR: The PIT framework for business re-engineering. In *The Proceedings of the 2nd International Strategic Information Systems Conference*, Barcelona, Spain, September.

Venkatraman, N. (1991) IT-induced Business Reconfiguration. In *The Corporation of the 1990s: Information Technology and Organisational Transformation*, Scott-Morton, M. (ed.), Oxford University Press, Oxford.

Wang, S. (1994) OO Modelling of Business Processes: Object-oriented Systems Analysis, *Information Systems Management*, pp.36-43.

Wastell, G. W., White, P. and Kawalek, P. (1994) A methodology for business process redesign: experience and issues, *Journal of Strategic Information Systems*, vol.3, no.1, pp.5-22.

Willcocks, L. and Smith, G. (1995) IT-enabled Business Process Reengineering: From Theory to Practice, In *The Proceedings of the 3rd European Conference on Information Systems*, Athens, Greece, pp.471-485.